AF413079

SPECTRAL TECHNIQUES IN VLSI CAD

Spectral Techniques in VLSI CAD

by

Mitchell Aaron Thornton
Mississippi State University

Rolf Drechsler
Siemens AG

and

D. Michael Miller
University of Victoria

KLUWER ACADEMIC PUBLISHERS

BOSTON / DORDRECHT / LONDON

A C.I.P. Catalogue record for this book is available from the Library of Congress.

ISBN 0-7923-7433-9

Published by Kluwer Academic Publishers,
P.O. Box 17, 3300 AA Dordrecht, The Netherlands.

Sold and distributed in North, Central and South America
by Kluwer Academic Publishers,
101 Philip Drive, Norwell, MA 02061, U.S.A.

In all other countries, sold and distributed
by Kluwer Academic Publishers,
P.O. Box 322, 3300 AH Dordrecht, The Netherlands.

Printed on acid-free paper

Printed in the Netherlands.

To

Misty, Micah, Henry

and

Monty, Dana

and

Judy

CONTENTS

Contents ix

PREFACE

The application of spectral techniques in digital systems design is not new and many results have been presented in conference proceedings and archival journal papers. There are also several excellent books on this and related subjects. Nevertheless, a central reference volume fully devoted to developments in the use of spectral techniques in conjunction with the now popularly used *Decision Diagram* (DD) data structure has previously not been available. Our objective in writing this book has been to provide such a reference and, in particular, to bring together results on the use of DDs in the computation, representation, analysis and application of a variety of Boolean function spectra.

DDs have revolutionized many areas of digital circuit design and are the basis for a multitude of new algorithmic and *Very Large Scale Integration* (VLSI) *Computer-Aided Design* (CAD) methods. A common phenomenon is that CAD algorithms that are reformulated to use DD data structures often result in techniques with practical computational resource requirements. Many times these methods were previously computationally limited and thus could not be applied to the size of problems encountered in the VLSI circuit design arena. A major theme of this book is to show that DDs bring a level of practicality to spectral techniques that, without their use, has not previously been available. The intended audience for this book is CAD tool developers and, users, graduate students and researchers with an interest in the practical application of spectral techniques to VLSI and digital systems design.

In the past VLSI CAD practitioners may not have found interest in spectral methods due to the typical presentation of these techniques where heavy reliance on linear algebra and other mathematical constructs is included. One of our main goals in writing this book has been to present spectral methods in a manner that is easy to understand in the context of DDs and DD-based algorithms. Hence, this book is not intended to present a formal theoretical basis for these techniques. Rather we have chosen to include theory only to the level necessary to understand the applications of the practical methods presented.

The book begins with a description of necessary background information on Boolean functions, spectral techniques and DDs. The computation of a variety of spectra of Boolean functions is considered using matrix methods, so-called *fast transform* techniques, and DD-based methods derived from those traditional approaches. The interrelations between various spectral domains and the Boolean domain are given.

The vitally important area of DD minimization is examined in detail with emphasis on spectral DDs and on expediting spectral computations. A summary of some methods for automated synthesis using spectral techniques is presented and the application of spectral methods to the verification problem is described.

It is our hope that this work will shed new light on the practicality of spectral techniques in VLSI CAD and will both promote the use of these techniques and inspire others to pursue further research in spectral methods.

Acknowledgement

Although many of the techniques presented here are the results of research endeavors by the authors, we have also included research results from the literature that we felt were particularly relevant. We are greatly indebted to the community of spectral technique and DD researchers, both past and present, and we regret that we may have overlooked some pertinent information that is present in the vast amount of literature in these areas.

We acknowledge the support of our respective institutions, Mississippi State University, Albert-Ludwigs-University, Siemens AG and the University of Victoria, and of our many colleagues there and elsewhere. Many of the results presented here resulted from financial support from the National Science Foundation in the U.S.A., the DAAD (*Deutscher Akademischer Austauschdienst*) in Germany and the Natural Sciences and Engineering Research Council of Canada. The help of James Finlay, Mark de Jongh and Cindy Lufting from Kluwer Academic Publishers was absolutely instrumental to the existence of this book and we are grateful. We also wish to thank Mary Ann Thornton and Whitney Townsend for proofreading the final manuscript.

Finally, we acknowledge our graduate students from whom we have learned much and, most importantly, the support and sacrifice provided by our families during the time this book was being written.

Spring 2001   *Mitchell Aaron Thornton*

Rolf Drechsler

D. Michael Miller

1

INTRODUCTION

Motivation for the consideration of the use of spectral methods in VLSI CAD methods is given, and a brief history of important research results is also included. The prerequisite mathematical background for easily understanding this book is outlined and this chapter concludes with a description of the organization of the remainder of the text.

1.1 MOTIVATION

Circuit design and analysis requires tools that support the description and manipulation of the circuit and its underlying functionality at various levels of abstraction ranging from the detailed physical device level through a high level systems specification. These tools typically employ mathematical models which are often of necessity simplified in order to provide easier manipulation and to permit the needed computation in a reasonable amount of time. Frequently, a variety of models are employed offering diverse views of the problem at hand.

A commonly used technique is to describe circuits based upon responses and characteristics in alternative functional spaces where basis functions are chosen to more closely resemble the circuit characteristics. For this type of analysis, it is necessary to transform circuit characteristics from one domain to another. Examples of this form of analysis are the set of complex, damped sinusoids that form the basis of the LaPlace transform and the set of harmonic undamped sinusoids that are used as a basis for the Fourier transform.

The unprecedented success of digital circuitry, especially binary-valued circuits, has led to a virtual revolution in terms of representing complex reasoning and underlying calculations in terms of electronics. Digital circuitry has led to a different set of design paradigms than those applied to other forms of electronic circuit design. For example, while most other electronic circuitry evolved using mathematical concepts including spectral transformations, most modern digital design techniques and tools are based upon the manipulation of discrete mathematical constructs and rely heavily on Boolean algebra. As an interesting corollary, the digitally based microprocessor has been employed to solve many problems in modern signal processing and has given rise to the now established subdiscipline of digital signal processing.

As technology matures, more transistors per square micron are being fabricated on *Integrated Circuits* (ICs). Modern complex VLSI devices contain millions of transistors. One enabling technology for this trend is the use of *Computer Aided Design* (CAD) software that allows for the automation of design tasks. A central issue in the development of CAD methods is the representation and manipulation of Boolean functions that model the underlying circuitry.

In the past there has been some interest in applying spectral methods to digital design problems. Although many interesting results, algorithms and techniques were devised, it soon became apparent that their use was greatly limited by the lack of techniques for computing and manipulating the spectra of large functions encountered in practical design situations.

In this book we consider a variety of spectral techniques that afford alternative views of Boolean functions and their implementations as digital circuits. Particular emphasis is placed on representations and methods that allow spectral techniques to be applied to larger problems than has earlier been possible.

1.2   BACKGROUND AND HISTORY

The mathematical underpinnings for the spectral techniques addressed in this book have a long history. In 1910 A. Haar [80] introduced a set of 2^n orthogonal functions that now bear his name. These functions form the basis for the discrete normalized Haar transform discussed in section 3.2.4. The more widely known Walsh transform, described in section 3.2.1, is based on a different set of 2^n discrete orthogonal functions introduced by J. L. Walsh in 1923 [167] based on work accomplished the previous year by H. Rademacher [132]. The

Walsh, Haar and a variety of other similar transforms have been widely applied to a variety of applications as documented in a number of books on the subject [1, 5, 90, 92, 97].

Two additional transforms are also considered in this book. The Reed-Muller transform, considered in section 3.2.2, is based on the 1954 work of I.S. Reed [133] and D.E. Muller [123]. The arithmetic transform, section 3.2.3, is a more recent development introduced by S. K. Kumar and M. A. Breuer in 1981 [104]. There is no intent to exhaust the vast choice of transforms. Rather the four cited are our focus and should be taken as representative.

Spectral techniques have been extensively considered with respect to function classification, logic design, testing and other applications in digital logic. Several books [5, 75, 92, 90, 97] provide excellent descriptions of this earlier work and while aspects of these are addressed in this book, the interested reader is encouraged to review them for a full understanding of the application of spectral techniques to the field of digital design. Although the theoretical background and techniques were very interesting, few successful practical applications were reported since the methods typically incurred exponential time and space complexities.

Decision diagrams have been of interest for many years to complexity theorists who refer to them as *branching programs*. In 1959 C.Y. Lee [109] published a paper describing the use of these constructs as data structures that could be used to represent switching functions in computer programs. This idea was later extended and studied in more depth by S.B. Akers [2] who coined the term *Binary Decision Diagrams* (BDDs). In 1986 the seminal paper by R.E. Bryant [15] introduced the notions of BDD reduction operations and variable ordering. These ideas resulted in a data structure that is canonical for a given Boolean function and allowed for manipulation algorithms to be developed that are quadratic in terms of the size of the BDD. Since then, extensive research on BDDs and other forms of *Decision Diagrams* (DDs) has occurred. References to this extensive body of work are provided in the bibliography.

Bryant's 1986 paper [15] led to the observation [34, 117, 71] that decision diagrams can be applied to the determination and representation of various spectra of Boolean functions. The major contributions of this book are to present a single reference on the application of decision diagram techniques to implement spectral methods and to demonstrate their practicality . Related results can be found in [137].

1.3 MATHEMATICAL REQUIREMENTS

We assume that the reader is familiar with the fundamentals of Boolean algebra and digital logic design as is found in standard undergraduate textbooks on these subjects. Spectral methods, DDs and their application to the representation of Boolean functions and the spectra of Boolean functions, are developed here from first principles. No prior background is required in those areas. Our treatment of spectral methods for digital logic and DDs is not exhaustive and we refer the interested reader to the vast literature on those subjects. An extensive bibliography is provided for this purpose.

The mathematics used in this book is straightforward with knowledge of basic matrix algebra and Boolean algebra being the only significant prerequisite. We do make extensive use of the *Kronecker product* of matrices, denoted $\otimes$ (this is also referred to as the direct or tensor product in other work). As this operation may not be fully familiar to all readers, we provide its definition and basic properties here without proofs. The latter may be found in [74].

Given a matrix $\mathbf{A}$ of order $(m \times n)$ with the element in the ith row and j^{th} column denoted by a_{ij} and a matrix $\mathbf{B}$ of order $(r \times s)$, the Kronecker product $\mathbf{A} \otimes \mathbf{B}$ is given by

$$
\mathbf{A} \otimes \mathbf{B} =
\begin{bmatrix}
a_{11}\mathbf{B} & a_{12}\mathbf{B} & \cdots & a_{1n}\mathbf{B} \\
a_{21}\mathbf{B} & a_{22}\mathbf{B} & \cdots & a_{2n}\mathbf{B} \\
\vdots & \vdots & & \vdots \\
a_{m1}\mathbf{B} & a_{m2}\mathbf{B} & \cdots & a_{mn}\mathbf{B}
\end{bmatrix}
$$

The product matrix has order $(mr \times ns)$. Note that unlike the normal matrix product, the Kronecker product is defined for any matrix order.

For matrices $\mathbf{A}$, $\mathbf{B}$, $\mathbf{C}$ and $\mathbf{D}$ and a scalar α, the following properties hold

$$
\begin{aligned}
(\alpha\mathbf{A}) \otimes \mathbf{B} &= \alpha(\mathbf{A} \otimes \mathbf{B}) \\
\mathbf{A} \otimes (\alpha\mathbf{B}) &= \alpha(\mathbf{A} \otimes \mathbf{B}) \\
(\mathbf{A} + \mathbf{B}) \otimes \mathbf{C} &= (\mathbf{A} \otimes \mathbf{C}) + (\mathbf{B} \otimes \mathbf{C}) \\
\mathbf{A} \otimes (\mathbf{B} + \mathbf{C}) &= (\mathbf{A} \otimes \mathbf{B}) + (\mathbf{A} \otimes \mathbf{C}) \\
\mathbf{A} \otimes (\mathbf{B} \otimes \mathbf{C}) &= (\mathbf{A} \otimes \mathbf{B}) \otimes \mathbf{C}
\end{aligned}
$$

$$\begin{aligned}
(\mathbf{A} \otimes \mathbf{B})^t &= \mathbf{A}^t \otimes \mathbf{B}^t \\
(\mathbf{A} \otimes \mathbf{B})^{-1} &= \mathbf{A}^{-1} \otimes \mathbf{B}^{-1} \\
(\mathbf{A} \otimes \mathbf{B})(\mathbf{C} \otimes \mathbf{D}) &= \mathbf{AC} \otimes \mathbf{BD}
\end{aligned} \qquad (1.1)$$

Equation 1.1 is termed the *mixed product rule* and is only valid when the matrices are of appropriate dimensions to allow for normal matrix multiplication. Finally, the following observation from the above properties is particularly relevant to our work; the Kronecker product of two symmetric matrices is itself a symmetric matrix.

1.4 ORGANIZATION

Chapter 2 introduces the basic notation and definitions used in this book. A brief survey of the mathematical definition of four spectral transforms and the notation used to represent them is given in Chapters 2 and 3. In particular, Chapter 3 also reviews the Cooley-Tukey type "fast" algorithms [35] for spectral computation and introduces some "fast" algorithms for the direct transformation from one spectral domain to another. Emphasis is placed on the four spectral transformations that are used throughout this book, the Walsh, Reed-Muller, Haar and arithmetic types.

This book utilizes modern representation and manipulation methods of Boolean functions for the computation and use of spectral methods. In particular, spectral transform computations for large functions will be examined using modern CAD data structures such as DDs and cube lists. Cube lists and output probability calculations are reviewed in Chapter 2, and Chapter 4 introduces various DD types with an emphasis on spectral interpretation.

Chapter 5 is devoted to new algorithms for the calculation of the spectrum using cube lists and graph-based data structures. These methods form the basis for practical application of spectral methods in the CAD area.

Although DDs have been used to solve many important problems in CAD, the well-known minimization problem must be considered. In Chapter 6, DD minimization is outlined and specific techniques based on spectral methods are described.

Next, a discussion of the application of spectral methods in VLSI CAD is presented in Chapters 7 and 8. The important areas of automated logic synthesis and verification are examined and descriptions of how problems in these areas may be solved using spectral methods are given. Chapter 9 concludes this book with a summary of the major contributions.

2

THE BOOLEAN DOMAIN

Definitions and notation used throughout the remainder of this book are introduced here. In addition, basic concepts of Boolean functions as used throughout this book are also reviewed. Output probabilities which are used extensively in later chapters are introduced and results relating them to Boolean functions are also presented.

2.1 BOOLEAN FUNCTIONS

We assume that the reader is familiar with the basic concepts of Boolean algebra and switching theory and here only touch upon the characteristics of Boolean functions that are most important for the work that follows.

Boolean variables are denoted x, y, z, *etc.* and take values from the set $B = \{0, 1\}$. Variables are frequently taken from a set $X = \{x_1, x_2, \ldots, x_n\}$.

The Boolean constants 0 and 1 are themselves simple Boolean functions. The Boolean variable x_i is a further example of a Boolean function $x_i : B^n \to B$ as is its complement $\overline{x}_i$. For a Boolean variable x_i we also use x_i^1 to denote x_i and x_i^0 to denote $\overline{x}_i$. The variables x_i^1 and x_i^0 are called Boolean *literals* of the variable x_i. x_i^1 represents a variable with positive polarity, while x_i^0 represents a variable with negative polarity.

A product (logical AND) of literals is called a *cube*. Logic-1 is the identity cube, that is, the cube which is the product of zero literals. Conversely, logic-0 is the cube resulting from the product of all possible literals. A cube which is the product of n literals each involving a different x_i from X is termed a minterm. In particular, for $a = (a_1, \ldots, a_n) \in B^n$,

$$m_a = x_1^{a_1} \cdot \ldots \cdot x_n^{a_n}$$

is the minterm corresponding to a.

A Boolean function can be specified by enumerating its value for each possible variable assignment, as is done in a truth table representation. Alternatively, for fully specified functions, only the minterms for which the function takes the value 1 need be listed; the others correspond to 0 by default. A cube corresponds to a set of minterms the members of which are found by assigning the variables not appearing in the cube to 0 and 1 in all possible ways. A Boolean function can thus be specified by a set of cubes where each minterm for which the function takes the value 1 is covered by at least one cube and no cube covers a minterm for which the function is 0. Such a cube set is termed a cube cover of the function.

A cube cover corresponds to the familiar *sum-of-products* representation in Boolean algebra with each cube corresponding to a product term and the function being the sum (logical OR) of those terms. Note that a single cube represents a Boolean function on its own. The set of cubes representing a Boolean function is clearly not necessarily unique. A case of interest in this book is when each pair of cubes in the cover is disjoint, that is, when the two cubes do not share a common minterm. Such a situation corresponds to a *Disjoint Sum-Of-Products* (DSOP) expression.

Cubes are most commonly written in a positional notation with each cube consisting of an n-tuple of symbols where a 0 in position i indicates $\overline{x}_i$ is in the cube, a 1 in position i indicates x_i appears in the cube, and a $-$ in position i indicates the cube does not involve x_i or $\overline{x}_i$.

2.1.1   Incompletely-Specified Boolean Functions

Quite often, *incompletely-specified* Boolean functions are of interest. Formally, an incompletely-specified Boolean function is a mapping

$$f : D \to B$$

with $D \subset B^n$. The minterms for which the value of the function is not specified are called *don't-care* conditions. An incompletely-specified Boolean function partitions the set of minterms into three disjoint sets: the first for which the function value is 1, the second for which the function value is 0 and the third for which the function value is a don't-care. These are termed the off-set ($OFF(f)$), the on-set ($ON(f)$) and the don't-care set ($DC(f)$), respectively. Any pair of these three sets uniquely defines the function, as the third set consists of all minterms not included in the other two. It is most common to use $ON(f)$ with $OFF(f)$ or $ON(f)$ with $DC(f)$ for this purpose. Clearly, each of the ON, OFF and DC-sets can be covered by sets of cubes.

From the above definitions, it is seen that an incompletely-specified Boolean function f can be uniquely defined by a pair of completely-specified functions. For example, let f_{ON} be the completely-specified function whose on-set is $ON(f)$ and whose off-set is $OFF(f) \cup DC(f)$. Furthermore, let f_{OFF} be the completely-specified function whose on-set is $OFF(f)$ and whose off-set is $ON(f) \cup DC(f)$. $f_{ON} = 1$ corresponds to $f = 1$, $f_{OFF} = 1$ corresponds to $f = 0$ and both being 0 corresponds to $f = $ don't-care. f_{DC} can be defined in a similar fashion to yield alternative representations for f.

In the following we assume all functions to be completely specified if not explicitly stated otherwise.

2.1.2 Cofactors and Function Decompositions

The two *cofactors* of f with respect to x_i are given by

$$f_{x_i=c}(x_1, .., x_{i-1}, x_i, x_{i+1}, .., x_n) = f(x_1, .., x_{i-1}, c, x_{i+1}, .., x_n)$$

with $c = 0$ or 1. The negative cofactor ($c = 0$) is denoted $f_{\overline{x}_i}$ or alternatively f_i^0. The positive cofactor ($c = 1$) is denoted f_{x_i} or f_i^1.

Since the Boolean difference (also commonly referred to as the Boolean derivative) is also used, we introduce

$$\frac{\partial f}{\partial x_i} = f_i^2 = f_i^1 \oplus f_i^0$$

where the partial derivative notation denotes the Boolean difference with respect to the dependent variable x_i. If it is clear from the context in which variable x_i is considered, we write f_0, f_1 and f_2 for convenience.

The cofactors and the Boolean difference of an n-variable Boolean function f are each functions of $n - 1$ variables.

The following theorem identifies three decompositions of a Boolean function.

Theorem 2.1 *Let f be a Boolean function defined over the set of variables, $\{x_1, x_2, \ldots, x_n\}$. For all $i \in \{1, \ldots, n\}$ it holds that*

$$
\begin{aligned}
f &= \overline{x}_i f_i^0 + x_i f_i^1 & \text{Shannon } (S) & \quad\quad (2.1) \\
f &= f_i^0 \oplus x_i f_i^2 & \text{positive Davio } (pD) & \quad\quad (2.2) \\
f &= f_i^1 \oplus \overline{x}_i f_i^2 & \text{negative Davio } (nD) & \quad\quad (2.3)
\end{aligned}
$$

The proof of the theorem is straightforward. Note that in Equation 2.1 the inclusive OR ($+$) can be replaced by an exclusive OR ($\oplus$). In addition, observe that the subfunctions identified in Equations 2.1-2.3 are uniquely determined.

2.1.3 Multiple-output Functions

Often one wishes to deal with a set of m Boolean functions over a common set of n Boolean variables. Such a situation can be defined as a single mapping $f : B^n \to B^m$. While this mapping correctly defines a system of Boolean functions, f is often referred to as a *multiple-output function* for convenience and we adopt that terminology here.

The concepts discussed above generally apply to multiple-output functions. When cubes are used to define a multiple-output function, each cube has an associated output tag which is an m-tuple where the coordinate corresponding to each output is 0, 1 or - indicating that the output is 0, 1 or a don't-care for every minterm covered by the cube.

2.2   PSEUDO-BOOLEAN FUNCTIONS

A *pseudo-Boolean function* is a mapping $f : B^n \to Z^m$ where Z is the set of integers. Note that $m = 1$ for a single-output function. The above observations for Boolean functions generally hold for pseudo-Boolean functions with the exception that the difference with respect to x_i is given by $f_i^2 = f_i^1 - f_i^0$.

Pseudo-Boolean functions can be decomposed in a manner similar to Boolean functions as given in the following theorem.

Theorem 2.2 *Let $f : \mathrm{B}^n \to \mathrm{Z}$ be an integer-valued function defined over the set of Boolean variables $\{x_1, x_2, \ldots, x_n\}$. For all $i \in \{1, \ldots, n\}$ it holds that*

$$
\begin{aligned}
f &= (1 - x_i)f_i^0 + x_i f_i^1 & Shannon\ (S) & \quad (2.4)\\
f &= f_i^0 + x_i f_i^2 & positive\ Davio\ (pD) & \quad (2.5)\\
f &= f_i^1 + (1 - x_i)(-f_i^2) & negative\ Davio\ (nD) & \quad (2.6)
\end{aligned}
$$

2.3　OUTPUT PROBABILITY

Output probabilities were proposed by Parker and McCluskey in [128] to evaluate the effectiveness of random testing in combinational logic circuits. Recently, interest has been renewed in these quantities since they can be used to form estimates of switching activity factors under some assumptions. Switching activity factors can be very useful in the prediction of overall power dissipation for CMOS circuitry and are thus a quantity to be minimized for low power design [73, 129, 163]. Later chapters in this book show other uses of output probabilities particularly with regard to spectral computations.

The output probability expression for a Boolean function is a real-valued algebraic equation that specifies the probability that the function is valued at logic-1 given the probabilities that each of the dependent variables are valued at logic-1. Therefore, the probability space consists of 2^n experiments where n is the number of dependent Boolean variables. If it is assumed that the function is completely specified, and that each input is equally likely to be 0 or 1, all probabilities for function variables may be set to $\frac{1}{2}$. The resulting circuit output probability will always have a value in the interval $[0, 1]$. This resulting probability is simply the fraction of minterms that cause the function to evaluate to logic-1 and this quantity is denoted as $\wp\{f\}$.

As an example, consider the function of six variables defined by the Boolean expression in Equation 2.7.

$$f(x) = x_1 x_3 \overline{x}_6 + x_1 \overline{x}_3 x_4 \overline{x}_6 + x_1 \overline{x}_3 \overline{x}_4 \overline{x}_5 + \overline{x}_1 x_2 x_4 \overline{x}_6 + \overline{x}_1 x_2 \overline{x}_4 \overline{x}_5 + x_1 \overline{x}_2 \overline{x}_5 \quad (2.7)$$

By simple enumeration, we find $\wp\{f\} = 0.421875$ since 27 of the possible 2^6 minterms cause f to evaluate to a logic-1 value.

In the work [128] two methods were presented for the computation of the output probability expressions for a logic circuit. The first method uses algebraic expressions and the second method uses a schematic diagram. Later, new methods were developed for the computation of these quantities using decision diagrams [24, 157].

2.3.1 Computation of Output Probabilities Using DSOP Cubes

When the set of covering cubes is in DSOP form, it is guaranteed that each minterm in the set $ON(f)$ is covered exactly once. Based on this observation, the quantity $\wp\{f\}$ is readily computed by summing the number of minterms covered by each cube in the DSOP cube list and dividing the sum by 2^n where n is the number of dependent variables of f.

Example 2.1 *Consider the function $f = \overline{x}_1 \overline{x}_3 + x_2 \overline{x}_3 + x_1 \overline{x}_2 x_3$. This function is written in SOP form with non-disjoint cubes. When f is expanded into a DSOP form, it may be written as $f = \overline{x}_1 \overline{x}_3 + x_1 x_2 \overline{x}_3 + x_1 \overline{x}_2 x_3$ (as is always the case with DSOP forms the + operator may be interchanged with the $\oplus$ operator). Expressed in DSOP form, we can write the description of $ON(f)$ as shown in Table 2.1. Using the method outlined above, $\wp\{f\} = (2 + 1 + 1)/8 = 1/2$. Therefore, it is easily seen that the output probability of a function may be computed by processing each cube in a DSOP list.*

2.3.2 Conditional Output Probabilities

The use of circuit output probabilities in digital systems engineering tasks in the past has generally centered around the computation of the probabilities

Table 2.1   Example of $ON(f)$ Described Using a DSOP Cube List

x_1	x_2	x_3	$N_{minterms}$
0	*-*	*0*	*2*
1	*1*	*0*	*1*
1	*0*	*1*	*1*

of single functions. In this section we use conditional output probability expressions to show how some interesting relationships may be derived and we show how output probability values may be related to cover sets of Boolean functions. Unless noted otherwise, the conditioned quantity will always be dependent upon the function of interest. Therefore, the probability space for the conditional output probability, denoted $\wp\{f|f_c\}$, has a size of 2^{n-m} where n is the number of dependent variables of f and m is the number of mutually dependent variables of f and f_c. f_c denotes a constituent function and serves as the conditional function in $\wp\{f|f_c\}$.

2.3.3   Output Probabilities of Shannon CoFactors

It is convenient to describe conditional output probabilities by examining the output probabilities of cofactor functions as used in the Shannon expansion, Equation 2.1. Many other useful operations such as the Boolean derivative, consensus and smoothing can be defined in terms of the cofactors f_0 and f_1.

Consider the cofactor f_1 for a particular variable x_i. The output probability is computed as the total number of times $f_1 = 1$ (this is commonly known as the *satisfy-count* of f_1) divided by 2^{n-1} if f is a completely specified function of n variables. In terms of probability theory, this becomes the conditional probability value $\wp\{f|x_i\}$.

Using Bayes' Rule a conditional probability may be expressed as given in Equation 2.8. Note that $\wp\{x_i\} \neq 0$ is always necessarily true when Equation 2.8 is used since the cofactors are always computed in terms of some dependent x_i. The case where $x_i = 0$ and hence $\wp\{x_i\} = 0$ indicates that f does not depend upon x_i and is thus not of interest.

$$\wp\{f|x_i\} = \frac{\wp\{f \cdot x_i\}}{\wp\{x_i\}} \qquad (2.8)$$

Since it is typical to deal with fully specified functions, $\wp\{x_i\} = \frac{1}{2}$ and Equation 2.8 simplifies to $\wp\{f|x_i\} = 2 \cdot \wp\{f \cdot x_i\}$.

2.3.4 Output Probabilities and Covers of Boolean Functions

It is possible to compute conditional probabilities for more complex constituent functions $\wp\{f|f_c\}$. The use of Bayes' rule allows a very succinct method for determining if a given function f covers another function f_c. Lemma 2.1 states and proves this result.

Lemma 2.1 *A function f covers another function f_c if, and only if,* $\wp\{f_c \cdot f\} = \wp\{f\}$.

Proof: By definition, a cover of function f_c by another function f means that for all $f_c = 1$, f is also at a logic-1 value. In terms of conditional probabilities, this is stated as $\wp\{f_c|f\} = 1$. Substituting this expression into Equation 2.8 yields the following result.

$$\wp\{f \cdot f_c\} = \wp\{f\} \qquad (2.9)$$

This substitution is valid for all cases where f and f_c have mutual dependence on a common subset of variables. This dependence assumption prevents $\wp\{f_c\} = 0$. It is trivial to ensure dependence since it is only necessary to show that at least one common variable is present in some canonical form of f and f_c. $\qquad\square$

This result can be useful in synthesis algorithms that require an efficient means for checking covers. Depending on the way f and f_c are represented initially, computing the two probability values in Equation 2.9 may prove to be much faster than other methods such as checking for intersections using cube sets.

Many logic optimization algorithms rely upon repeatedly computing relationships among a set of cubes representing a logic function. In particular, the computations commonly involve determining if cubes are disjoint, totally redundant or partially redundant. Conditional probability values may be used to quickly determine which of these three relationships are present between any pair of cubes from a set that represent some function f.

The *totally redundant* category is really a special case of Lemma 2.1. In this case we say that a cube c_i is totally redundant with respect to a cube c_j if c_j covers c_i. Two cubes c_i and c_j are *disjoint* when they each cover mutually exclusive portions of the range of the Boolean function f. Finally, c_i and c_j are *partially redundant* if they each cover an intersecting subset of the range of f. The following lemmas show the relationships between the conditional probabilities among pairs of cubes.

Lemma 2.2 *Given two distinct cubes c_i and c_j from a cover set of a function f, $\wp\{c_i|c_j\} = 1$ if, and only if, c_i is totally redundant.*

Proof: Since $\wp\{c_j\} \neq 0$, the conditional probability can be computed as given in Equation 2.10.

$$\wp\{c_i|c_j\} = \frac{\wp\{c_i \cdot c_j\}}{\wp\{c_j\}} \tag{2.10}$$

When cube c_i is totally redundant, $c_i \cdot c_j = c_j$. Substituting this result into Equation 2.10 it is seen that $\wp\{c_i|c_j\} = 1$. $\square$

Lemma 2.3 *Given two distinct cubes c_i and c_j from a cover set of a function f, $\wp\{c_i|c_j\} = 0$ if, and only if, c_i and c_j are disjoint.*

Proof: Since $\wp\{c_j\} \neq 0$, the conditional probability can be computed as follows.

$$\wp\{c_i|c_j\} = \frac{\wp\{c_i \cdot c_j\}}{\wp\{c_j\}} \tag{2.11}$$

When the cubes c_i and c_j are disjoint, $c_i \cdot c_j = 0$. Thus, the conditional probability becomes $\wp\{c_i|c_j\} = \frac{\wp\{0\}}{\wp\{c_j\}} = 0$. $\qquad\qquad\square$

Lemma 2.4 *Given two cubes c_i and c_j from a cover set of a function f, $\wp\{c_i|c_j\} = p$, where p lies in the interval $(0,1)$ if, and only if, c_i and c_j are partially redundant.*

Proof: When $\wp\{c_i|c_j\} = p$, there is necessarily some dependence between c_i and c_j, therefore they cannot be disjoint and $\wp\{c_i|c_j\} \neq 0$. Further, since $p < 1$, c_i is not totally redundant. The only possible remaining relationship between c_i and c_j is that they are partially redundant. $\qquad\square$

When a cube is a specific minterm, m_i, the conditional probability becomes $\wp\{f|m_i\} = 0$ or $\wp\{f|m_i\} = 1$. This is due to the fact that the probability space is reduced in size to $2^{n-n} = 1$, hence it consists of a single experiment. The outcome of this experiment is that the function either covers the minterm or does not. Hence, a conditional probability $\wp\{f|m_i\}$ indicates whether the function depends upon m_i and that the real value of the probability is the same as the Boolean value of the function when it is evaluated at that particular minterm. This result comes directly from the above lemmas since it is impossible for a specific minterm and a function to be partially redundant.

2.3.5 Smoothing, Consensus, and the Boolean Derivative

Three other operators are important in Boolean algebra and may be defined in terms of cofactors. These are the Boolean derivative (Boolean difference) Equation 2.12, the consensus operator (universal quantifier, $\forall$) Equation 2.13, and the smoothing operator (existential quantifier, $\exists$) Equation 2.14. These operations are interrelated in the Boolean domain as expressed in Equation 2.15.

$$\frac{\partial f}{\partial x_i} = f_0 \oplus f_1 \qquad\qquad (2.12)$$

$$C_{x_i}(f) = f_0 \cdot f_1 \tag{2.13}$$

$$S_{x_i}(f) = f_0 + f_1 \tag{2.14}$$

$$\overline{C}_{x_i}(f) \cdot S_{x_i}(f) = \frac{\partial f}{\partial x_i} \tag{2.15}$$

The Boolean derivative provides a measure of the observability of a particular input variable. When $\partial f / \partial x_i = 0$, x_i is an unobservable input. Therefore, the observability may be quantified by computing the output probability of $\partial f / \partial x_i$ with values close to zero indicating that x_i is relatively unobservable and does not greatly affect the two cofactors in different ways.

The consensus of a Boolean function represents the component of the function that is independent of a variable x_i. Thus, very small output probabilities of the consensus function can be used to indicate that some function f is highly dependent upon a variable x_i.

The smoothing operator represents the behavior of a function when the dependence of that function on a particular variable x_i is suppressed. Smoothing can also be used as a measure of the degree that a particular function f depends upon a variable x_i.

In terms of the computation of the output probabilities of these functions, it is only necessary to compute the four probabilities, $\wp\{f_1 \cdot f_0\}$, $\wp\{f_1 \cdot \overline{f}_0\}$, $\wp\{\overline{f}_1 \cdot f_0\}$, and $\wp\{\overline{f}_1 \cdot \overline{f}_0\}$ due to the following relationships:

$$\wp\{\frac{\partial f}{\partial x_i}\} = \wp\{f_1 \cdot \overline{f}_0\} + \wp\{\overline{f}_1 \cdot f_0\} - \wp\{f_1 \cdot \overline{f}_0\}\wp\{\overline{f}_1 \cdot f_0\} \tag{2.16}$$

$$\wp\{C_{x_i}(f)\} = \wp\{f_1 \cdot f_0\} \tag{2.17}$$

$$\wp\{S_{x_i}(f)\} = 1 - \wp\{\overline{f}_1 \cdot \overline{f}_0\} \qquad (2.18)$$

2.4 SUMMARY

This chapter has introduced the basic background on Boolean functions used throughout the remainder of this work. Readers requiring more background on Boolean algebra and functions should consult any of the many standard texts on the subject [84, 87, 101, 114, 139]. Output probability and its relation to Boolean function concepts was also considered and was related to some important operations in the Boolean domain.

3

THE SPECTRAL DOMAIN

The fundamental structures and properties of spectral transforms and the resulting spectra of Boolean functions is presented here. We restrict ourselves to the mathematical foundations necessary for the development and understanding of the techniques introduced in later chapters. Those wishing a more in-depth and rigorous mathematical treatment of the subject should consult the literature [1, 5, 97, 90, 92]. Matrix based computation of the spectra is considered as are "fast" transform techniques derived from the matrix structures. Alternative spectral computation methods, in particular those based on DDs, are discussed in Chapter 5.

3.1 SPECTRA OF LOGIC FUNCTIONS

A function can be transformed from the Boolean domain to a number of alternative spectral domains. We begin with the basic principles and at this point consider only completely-specified single-output functions. Broadening this viewpoint to consider the incompletely-specified and multiple-output cases will be considered in later chapters.

An n-input completely-specified Boolean function f can be represented by a column vector $\mathbf{Y} = \{m_0, m_1, m_2 \ldots m_{2^n-1}\}^T$ with 2^n entries each giving the functional value for the corresponding minterm. $\mathbf{Y}$ is often called the truth vector for f.

A Boolean function f represented by $\mathbf{Y}$ can be transformed from the Boolean to a spectral domain as shown in the following

$$\mathbf{R} = \mathbf{T}^n \mathbf{Y} \tag{3.1}$$

where $\mathbf{T}^n$ is a 2^n by 2^n transform matrix the precise specification of which defines the spectral domain in question. In certain cases the matrix has a simple recursive structure which can be used to significant computational advantage as will be shown.

Properties of the spectra and related transform matrices are examined later in this chapter. It is noted that we restrict our interest to invertible transforms, hence

$$\mathbf{Y} = (\mathbf{T}^n)^{-1} \mathbf{R}$$

holds for the transforms described here. The consequence of this is that while the transforms between the Boolean and spectral domains fully preserve information, but the spectral domains make certain properties easier to consider than in the Boolean domain, and different spectral domains illuminate different functional properties.

Often the transform matrix can be expressed as a sequence of Kronecker products of a single base matrix. The following theorem concerning inverses of such matrices is thus very useful.

Theorem 3.1 *Given a square invertible matrix* $\mathbf{A}$,

$$\left[\bigotimes_{i=1}^{n} \mathbf{A} \right]^{-1} = \bigotimes_{i=1}^{n} \mathbf{A}^{-1}$$

Proof: This follows immediately by the iterative application of the identity $(\mathbf{A} \otimes \mathbf{B})^{-1} = \mathbf{A}^{-1} \otimes \mathbf{B}^{-1}$ and the associativity of the Kronecker product. $\square$

See Section 1.3 for a definition and review of the properties of the Kronecker product. Full details may be found in [74]. Although not stated in [74], it is clear from the development there that the Kronecker product properties we make use of in this chapter hold over $GF(2)$, a fact that will be useful below in the consideration of the Reed-Muller transform.

For several of the transforms considered here, each row of the transform matrix is associated with a particular subset of $\{x_1, x_2, \ldots, x_n\}$, and the corresponding spectral coefficients will be indexed by those variables. Hence, in such cases we have

$$\mathbf{R} = \{r_0, r_1 \ldots r_n, r_{12}, r_{13} \ldots r_{n-1,n}, r_{123} \ldots r_{1\ldots n}\}$$

where r_0 is a notational convenience with 0 denoting the set of no variables, that is, the coefficient is related to a constant function.

It is necessary on occasion to represent a partition of a function or spectrum vector which is indicated by superscripts, *e.g.*

$$\mathbf{R} = \begin{bmatrix} \mathbf{R}^0 \\ \mathbf{R}^1 \end{bmatrix} \quad \mathbf{R} = \begin{bmatrix} \mathbf{R}^0 \\ \mathbf{R}^1 \\ \mathbf{R}^2 \\ \mathbf{R}^3 \end{bmatrix}$$

The elements of the partition have equal sizes.

3.2 SPECTRAL TRANSFORMS

Four particular spectral transforms that have been extensively studied in the literature are presented in this section: the Walsh, the Reed-Muller, the arithmetic, and the Haar transforms.

3.2.1 Walsh Transform

Perhaps the most well known and most widely studied spectral transforms are based on a set of orthogonal functions defined by J. L. Walsh in 1923 [167] which are an extension of a set of functions defined by H. Rademacher [132] a year earlier. The transform itself is a form of Hadamard matrix [166].

The *Walsh transform* matrix $\mathbf{W}^n$ in Hadamard order [86] can be defined as:

$$\mathbf{W}^0 = \begin{bmatrix} 1 \end{bmatrix} \quad \mathbf{W}^n = \begin{bmatrix} \mathbf{W}^{n-1} & \mathbf{W}^{n-1} \\ \mathbf{W}^{n-1} & -\mathbf{W}^{n-1} \end{bmatrix}$$

An equivalent definition using the Kronecker product is particularly useful where

$$\mathbf{W}^1 = \begin{bmatrix} 1 & 1 \\ 1 & -1 \end{bmatrix}$$

and $\mathbf{W}^n = \mathbf{W}^1 \otimes \mathbf{W}^{n-1}$.

Since the Kronecker product is associative, this may be written as given in the following equation.

$$\mathbf{W}^n = \bigotimes_{i=1}^{n} \mathbf{W}^1$$

The rows of $\mathbf{W}^n$ are the set of 2^n n-variable Walsh functions [167] of which the n-variable Rademacher functions [132] are a subset. In addition to the Hadamard (Walsh-Hadamard or natural), the Walsh (Walsh-Kaczmarz or sequency), the Paley-Walsh (or dyadic), and the Rademacher-Walsh orderings have been studied [5, 10, 92]. The Hadamard ordering has seen the most use since the simple recursive structure of the transform matrix allows for "fast" transform methods [35, 146] (see Section 3.3). The Hadamard, Walsh and Paley-Walsh orderings share the very useful property that the transform matrix is its own inverse with a scaling factor of $\frac{1}{2^n}$ as will be shown below for the Hadamard case. The practical importance of this is that the same computational procedure can be used for transforming between the function and spectral domains with the simple adjustment of scaling.

The Walsh spectrum $\mathbf{R}$ of a Boolean function f is given by

$$\mathbf{R} = \mathbf{W}^n \mathbf{Y}$$

where the matrix multiplication is carried out over the integers, with logic-0(1) treated as the integer 0(1). This mapping is termed R-encoding.

Example 3.1 illustrates the computation of the Rademacher-Walsh spectrum by matrix multiplication. In general, this approach require 2^{2n} multiplications and 2^{2n} additions.

Example 3.1 *Computation of the Walsh spectrum of a 3-variable function using R-encoding.*

$$
\begin{bmatrix} 3 \\ -1 \\ -1 \\ -1 \\ 1 \\ 1 \\ 1 \\ -3 \end{bmatrix}
=
\begin{bmatrix}
1 & 1 & 1 & 1 & 1 & 1 & 1 & 1 \\
1 & -1 & 1 & -1 & 1 & -1 & 1 & -1 \\
1 & 1 & -1 & -1 & 1 & 1 & -1 & -1 \\
1 & -1 & -1 & 1 & 1 & -1 & -1 & 1 \\
1 & 1 & 1 & 1 & -1 & -1 & -1 & -1 \\
1 & -1 & 1 & -1 & -1 & 1 & -1 & 1 \\
1 & 1 & -1 & -1 & -1 & -1 & 1 & 1 \\
1 & -1 & -1 & 1 & -1 & 1 & 1 & -1
\end{bmatrix}
\begin{bmatrix} 0 \\ 1 \\ 1 \\ 0 \\ 0 \\ 0 \\ 0 \\ 1 \end{bmatrix}
$$

An alternate formulation represents the function by the vector $\mathbf{Z}$ in which logic 0 is coded as $+1$ and logic 1 is coded as -1, which we term S-encoding. In this case the spectrum is given by

$$\mathbf{S} = \mathbf{W}^n \mathbf{Z}$$

Example 3.2 *Computation of the Rademacher-Walsh spectrum of a 3-variable function using S-encoding.*

$$
\begin{bmatrix} 2 \\ 2 \\ 2 \\ 2 \\ -2 \\ -2 \\ -2 \\ 6 \end{bmatrix}
=
\begin{bmatrix}
1 & 1 & 1 & 1 & 1 & 1 & 1 & 1 \\
1 & -1 & 1 & -1 & 1 & -1 & 1 & -1 \\
1 & 1 & -1 & -1 & 1 & 1 & -1 & -1 \\
1 & -1 & -1 & 1 & 1 & -1 & -1 & 1 \\
1 & 1 & 1 & 1 & -1 & -1 & -1 & -1 \\
1 & -1 & 1 & -1 & -1 & 1 & -1 & 1 \\
1 & 1 & -1 & -1 & -1 & -1 & 1 & 1 \\
1 & -1 & -1 & 1 & -1 & 1 & 1 & -1
\end{bmatrix}
\begin{bmatrix} 1 \\ -1 \\ -1 \\ 1 \\ 1 \\ 1 \\ 1 \\ -1 \end{bmatrix}
$$

Since $\mathbf{Z} = \mathbf{1} - 2\mathbf{Y}$, it can be shown that

$$s_0 = 2^n - 2r_0; \quad s_\alpha = -2r_\alpha, \forall\, \alpha \subseteq \{1, 2 \ldots n\}, \alpha \neq \emptyset \tag{3.2}$$

so the information content of the $\mathbf{R}$ and the $\mathbf{S}$ spectral coefficients is the same.

Theorem 3.2 $(\mathbf{W}^n)^{-1} = \frac{1}{2^n}\mathbf{W}^n.$

Proof: The proof follows from Theorem 3.1, the fact that $(\mathbf{W}^1)^{-1} = \frac{1}{2}\mathbf{W}^1$ and the fact that scalar multipliers can be factored out of a Kronecker product. $\square$

3.2.2 Reed-Muller Transform

The *Reed-Muller transform* is motivated by the seminal work in 1954 of I.S. Reed [133] and D.E. Muller [123] which led to considerable interest in the Reed-Muller (AND-XOR) expansion of Boolean functions. The transform matrix $\mathbf{M}^n$ is defined by

$$\mathbf{M}^0 = [\ 1\ ] \quad \mathbf{M}^n = \begin{bmatrix} \mathbf{M}^{n-1} & 0 \\ \mathbf{M}^{n-1} & \mathbf{M}^{n-1} \end{bmatrix} \tag{3.3}$$

and the spectrum $\mathbf{R}$ is given by

$$\mathbf{R} = \mathbf{M}^n \mathbf{Y} \tag{3.4}$$

In this case the matrix multiplication is over $GF(2)$, *i.e.* integer addition is replaced with summation (*mod* 2). $\mathbf{M}^n$ can be expressed using the Kronecker product as

$$\mathbf{M}^1 = \begin{bmatrix} 1 & 0 \\ 1 & 1 \end{bmatrix}$$

$$\mathbf{M}^n = \bigotimes_{i=1}^{n} \mathbf{M}^1 \tag{3.5}$$

Example 3.3 *Computation of the Reed-Muller spectrum of a 3-variable function.*

$$
\begin{bmatrix} 0 \\ 1 \\ 1 \\ 0 \\ 0 \\ 1 \\ 1 \\ 1 \end{bmatrix} = \begin{bmatrix} 1 & 0 & 0 & 0 & 0 & 0 & 0 & 0 \\ 1 & 1 & 0 & 0 & 0 & 0 & 0 & 0 \\ 1 & 0 & 1 & 0 & 0 & 0 & 0 & 0 \\ 1 & 1 & 1 & 1 & 0 & 0 & 0 & 0 \\ 1 & 0 & 0 & 0 & 1 & 0 & 0 & 0 \\ 1 & 1 & 0 & 0 & 1 & 1 & 0 & 0 \\ 1 & 0 & 1 & 0 & 1 & 0 & 1 & 0 \\ 1 & 1 & 1 & 1 & 1 & 1 & 1 & 1 \end{bmatrix} \begin{bmatrix} 0 \\ 1 \\ 1 \\ 0 \\ 0 \\ 0 \\ 0 \\ 1 \end{bmatrix}
$$

Theorem 3.3 $(\mathbf{M}^n)^{-1} = \mathbf{M}^n$ *over* $GF(2)$.

Proof: The proof follows from Theorem 3.1 and the fact $(\mathbf{M}^1)^{-1} = \mathbf{M}^1$ over $GF(2)$. $\qquad\square$

From this theorem we have

$$\mathbf{Y} = \mathbf{M}^n \mathbf{R} \tag{3.6}$$

The above shows that $\mathbf{Y}$ is a linear combination (over $GF(2)$) of the columns of $\mathbf{M}^n$ for which the relevant coefficient in $\mathbf{R}$ is 1. Each column of $\mathbf{M}^n$ represents a function which is the logical AND of a subset of $x_1, x_2, \ldots, x_n$. The leftmost column is the constant function 1 which corresponds to the AND of no variables. Hence the Reed-Muller spectrum identifies a representation for a Boolean function as a sum over $GF(2)$ as a collection of products of variables. To be precise,

$$\mathbf{Y} = \bigoplus_{i=0}^{2^n-1} \mathbf{r}_i \mathbf{M}_i^n \tag{3.7}$$

where $\mathbf{M}_i^n$ is the ith column of $\mathbf{M}^n$.

3.2.3 Arithmetic Transform

The *arithmetic transform* [85], which is also known as the *probability transform* [165] and the *inverse integer Reed-Muller transform* [33], was initially introduced by S.K. Kumar and M.A. Breuer in 1981 [104] in work on probabilistic aspects of Boolean functions. The transform matrix has a recursive structure analogous to that of the Walsh and Reed-Muller transforms and is given by

$$\mathbf{A}^0 = [\; 1 \;] \quad \mathbf{A}^n = \begin{bmatrix} \mathbf{A}^{n-1} & 0 \\ -\mathbf{A}^{n-1} & \mathbf{A}^{n-1} \end{bmatrix} \tag{3.8}$$

or alternatively

$$\mathbf{A}^1 = \begin{bmatrix} 1 & 0 \\ -1 & 1 \end{bmatrix}$$

$$\mathbf{A}^n = \bigotimes_{i=1}^{n} \mathbf{A}^1 \tag{3.9}$$

As before, we define the spectrum as

$$\mathbf{R} = \mathbf{A}^n \mathbf{Y} \tag{3.10}$$

Theorem 3.4

$$(\mathbf{A}^n)^{-1} = \bigotimes_{i=1}^{n} \begin{bmatrix} 1 & 0 \\ 1 & 1 \end{bmatrix}$$

Proof: The proof follows from Theorem 3.1 and the fact $(\mathbf{A}^1)^{-1} = \begin{bmatrix} 1 & 0 \\ 1 & 1 \end{bmatrix}$.
$\square$

Note that while $(\mathbf{A}^1)^{-1} = \mathbf{M}^1$, their use is quite different since the arithmetic spectrum is computed over the integers; whereas, the Reed-Muller spectrum is computed over $GF(2)$. It is for this reason the arithmetic transform was termed the inverse integer Reed-Muller transform in [33].

3.2.4 Haar Transform

The orthogonal Haar functions presented by A. Haar in 1910 [80] form a set of 2^n continuous orthogonal functions over the interval $[0,1]$. They can be defined as follows:

$$
\begin{aligned}
H_0^0(k) &= +1.0 \\
H_i^q(k) &= (\sqrt{2})^{i-1}(+1.0), \quad \text{for } \frac{q}{2^{i-1}} \le k < \frac{q+\frac{1}{2}}{2^{i-1}} \\
&= (\sqrt{2})^{i-1}(-1.0), \quad \text{for } \frac{q+\frac{1}{2}}{2^{i-1}} \le k < \frac{q+1}{2^{i-1}} \\
&= 0, \quad \text{at all other points}
\end{aligned}
\tag{3.11}
$$

where k is over the continuous interval 0 to 1 and $i = 1, 2, \ldots, n$ and $q = 0, 1, \ldots, 2^{i-1} - 1$.

Discrete sampling of the set of Haar functions gives a $2^n \times 2^n$ orthogonal matrix $\mathbf{T}^n$. For $n = 3$,

$$
\mathbf{T}^3 =
\begin{bmatrix}
1 & 1 & 1 & 1 & 1 & 1 & 1 & 1 \\
1 & 1 & 1 & 1 & -1 & -1 & -1 & -1 \\
\sqrt{2} & \sqrt{2} & -\sqrt{2} & -\sqrt{2} & 0 & 0 & 0 & 0 \\
0 & 0 & 0 & 0 & \sqrt{2} & \sqrt{2} & -\sqrt{2} & -\sqrt{2} \\
2 & -2 & 0 & 0 & 0 & 0 & 0 & 0 \\
0 & 0 & 2 & -2 & 0 & 0 & 0 & 0 \\
0 & 0 & 0 & 0 & 2 & -2 & 0 & 0 \\
0 & 0 & 0 & 0 & 0 & 0 & 2 & -2
\end{bmatrix}
\tag{3.12}
$$

$\mathbf{T}^n$ is a complete, orthogonal matrix with $\Sigma_{k=0}^{2^n-1} t_{ik} t_{jk} = 2^n$ if $i = j$ and 0 otherwise. The following theorem is a direct result from this definition of $\mathbf{T}^n$.

Theorem 3.5 $[\mathbf{T}^n]^{-1} = \frac{1}{2^n}[\mathbf{T}^n]^t$.

Note that $\mathbf{T}^n$ is not symmetric, so the transpose is needed for the inverse.

A computationally more practical *normalized Haar transform* $\mathbf{K}^n$ is derived from $\mathbf{T}^n$ by replacing the nonzero entries of $\mathbf{T}^n$ with the values +1 and -1 depending on their arithmetic sign. For $n = 3$, this yields:

$$
\mathbf{K}^3 = \begin{bmatrix}
1 & 1 & 1 & 1 & 1 & 1 & 1 & 1 \\
1 & 1 & 1 & 1 & -1 & -1 & -1 & -1 \\
1 & 1 & -1 & -1 & 0 & 0 & 0 & 0 \\
0 & 0 & 0 & 0 & 1 & 1 & -1 & -1 \\
1 & -1 & 0 & 0 & 0 & 0 & 0 & 0 \\
0 & 0 & 1 & -1 & 0 & 0 & 0 & 0 \\
0 & 0 & 0 & 0 & 1 & -1 & 0 & 0 \\
0 & 0 & 0 & 0 & 0 & 0 & 1 & -1
\end{bmatrix}
$$

For the case of $n = 4$, the normalized Haar transform is

$$
\begin{bmatrix}
1 & 1 & 1 & 1 & 1 & 1 & 1 & 1 & 1 & 1 & 1 & 1 & 1 & 1 & 1 & 1 \\
1 & 1 & 1 & 1 & 1 & 1 & 1 & 1 & -1 & -1 & -1 & -1 & -1 & -1 & -1 & -1 \\
1 & 1 & 1 & 1 & -1 & -1 & -1 & -1 & 0 & 0 & 0 & 0 & 0 & 0 & 0 & 0 \\
0 & 0 & 0 & 0 & 0 & 0 & 0 & 0 & 1 & 1 & 1 & 1 & -1 & -1 & -1 & -1 \\
1 & 1 & -1 & -1 & 0 & 0 & 0 & 0 & 0 & 0 & 0 & 0 & 0 & 0 & 0 & 0 \\
0 & 0 & 0 & 0 & 1 & 1 & -1 & -1 & 0 & 0 & 0 & 0 & 0 & 0 & 0 & 0 \\
0 & 0 & 0 & 0 & 0 & 0 & 0 & 0 & 1 & 1 & -1 & -1 & 0 & 0 & 0 & 0 \\
0 & 0 & 0 & 0 & 0 & 0 & 0 & 0 & 0 & 0 & 0 & 0 & 1 & 1 & -1 & -1 \\
1 & -1 & 0 & 0 & 0 & 0 & 0 & 0 & 0 & 0 & 0 & 0 & 0 & 0 & 0 & 0 \\
0 & 0 & 1 & -1 & 0 & 0 & 0 & 0 & 0 & 0 & 0 & 0 & 0 & 0 & 0 & 0 \\
0 & 0 & 0 & 0 & 1 & -1 & 0 & 0 & 0 & 0 & 0 & 0 & 0 & 0 & 0 & 0 \\
0 & 0 & 0 & 0 & 0 & 0 & 1 & -1 & 0 & 0 & 0 & 0 & 0 & 0 & 0 & 0 \\
0 & 0 & 0 & 0 & 0 & 0 & 0 & 0 & 1 & -1 & 0 & 0 & 0 & 0 & 0 & 0 \\
0 & 0 & 0 & 0 & 0 & 0 & 0 & 0 & 0 & 0 & 1 & -1 & 0 & 0 & 0 & 0 \\
0 & 0 & 0 & 0 & 0 & 0 & 0 & 0 & 0 & 0 & 0 & 0 & 1 & -1 & 0 & 0 \\
0 & 0 & 0 & 0 & 0 & 0 & 0 & 0 & 0 & 0 & 0 & 0 & 0 & 0 & 1 & -1
\end{bmatrix}
$$

Example 3.4 *Computation of the Haar spectrum of a 3-variable function using R-encoding.*

$$
\begin{bmatrix}
3 \\ 1 \\ 0 \\ -1 \\ -1 \\ 1 \\ 0 \\ -1
\end{bmatrix}
=
\begin{bmatrix}
1 & 1 & 1 & 1 & 1 & 1 & 1 & 1 \\
1 & 1 & 1 & 1 & -1 & -1 & -1 & -1 \\
1 & 1 & -1 & -1 & 0 & 0 & 0 & 0 \\
0 & 0 & 0 & 0 & 1 & 1 & -1 & -1 \\
1 & -1 & 0 & 0 & 0 & 0 & 0 & 0 \\
0 & 0 & 1 & -1 & 0 & 0 & 0 & 0 \\
0 & 0 & 0 & 0 & 1 & -1 & 0 & 0 \\
0 & 0 & 0 & 0 & 0 & 0 & 1 & -1
\end{bmatrix}
\begin{bmatrix}
0 \\ 1 \\ 1 \\ 0 \\ 0 \\ 0 \\ 0 \\ 1
\end{bmatrix}
$$

Theorem 3.6 *The normalized Haar transform can be expressed as*

$$
\mathbf{K}^0 = \begin{bmatrix} 1 \end{bmatrix} \quad \mathbf{K}^n = \begin{bmatrix} \mathbf{K}^{n-1} \otimes \begin{bmatrix} 1 & 1 \end{bmatrix} \\ \mathbf{I}^{n-1} \otimes \begin{bmatrix} 1 & -1 \end{bmatrix} \end{bmatrix} \tag{3.13}
$$

Proof: For the normalized Haar transform, Equation 3.11 becomes

$$
\begin{aligned}
H_0^0(k) \;&=\; +1 \\[4pt]
H_i^q(k) \;&=\; +1, \quad \text{for } \frac{q}{2^{i-1}} \le k < \frac{q+\frac{1}{2}}{2^{i-1}} \\[6pt]
&=\; -1, \quad \text{for } \frac{q+\frac{1}{2}}{2^{i-1}} \le k < \frac{q+1}{2^{i-1}} \\[6pt]
&=\; 0, \quad \text{at all other points}
\end{aligned}
\tag{3.14}
$$

where $i = 1, 2, \ldots, n$ and $q = 0, 1, \ldots, 2^{i-1} - 1$.

For $i = n$, 2^{n-1} Haar functions are defined, each sampled at 2^n points which are q and $q + \frac{1}{2}$ for $q = 0, 1, \ldots, 2^{n-1} - 1$. The first of these functions $H_n^0(k)$ is a 1, followed by a -1, followed by $(2^n - 2)$ 0's. The second, $H_n^1(k)$, is two 0's, followed by a 1, followed by a -1, followed by $(2^n - 4)$, 0's. The ongoing pattern should be apparent and is illustrated above for the case of $n = 3$. These functions in order are the bottom 2^{n-1} rows of $\mathbf{K}^n$. They can be expressed in matrix form as $\mathbf{I}^{n-1} \otimes [1 \ \ -1]$.

For $i = 1, 2, \ldots, n - 1$, the Haar functions that are defined and preceded by $H_0^0(k)$ are precisely those that compose $\mathbf{K}^{n-1}$, and it is these functions that comprise the upper half of $\mathbf{K}^n$. The difference is that, in order to correspond to the lower half of $\mathbf{K}^n$, these functions must be sampled twice as often. This corresponds to duplicating the values across the function which can be expressed in matrix form as $\mathbf{K}^{n-1} \otimes [1 \ 1]$.

Concatenating the two matrix expressions yields Equation 3.13. $\qquad\square$

Theorem 3.5 does not hold for the normalized Haar transform matrix since, while the rows do maintain pairwise orthogonality, the resultant values are not the same. The inverse of $\mathbf{K}^n$ is given by the following theorem.

Theorem 3.7 $\mathbf{K}^0 = [1]$

$$
(\mathbf{K}^n)^{-1} = \frac{1}{2^n} \left[(\mathbf{K}^{n-1})^{-1} \otimes \begin{bmatrix} 1 \\ 1 \end{bmatrix}, \mathbf{I}^{n-1} \otimes \begin{bmatrix} 2^{n-1} \\ -2^{n-1} \end{bmatrix} \right]
$$

Proof: Let

$$\mathbf{B}^n = \left[(\mathbf{B}^{n-1}) \otimes \begin{bmatrix} 1 \\ 1 \end{bmatrix}, \mathbf{I}^{n-1} \otimes \begin{bmatrix} 2^{n-1} \\ -2^{n-1} \end{bmatrix} \right]$$

and consider $\mathbf{K}^n \mathbf{B}^n$. This yields
$\mathbf{K}^n \mathbf{B}^n =$

$$\left[\begin{array}{cc} \left(\mathbf{K}^{n-1} \otimes [1\ \ 1]\right) \left(\mathbf{B}^{n-1} \otimes \begin{bmatrix} 1 \\ 1 \end{bmatrix} \right), & \left(\mathbf{K}^{n-1} \otimes [1\ \ 1]\right) \left(\mathbf{I}^{n-1} \otimes \begin{bmatrix} 2^{n-1} \\ -2^{n-1} \end{bmatrix} \right) \\ \left(\mathbf{I}^{n-1} \otimes [1\ -1]\right) \left(\mathbf{B}^{n-1} \otimes \begin{bmatrix} 1 \\ 1 \end{bmatrix} \right), & \left(\mathbf{I}^{n-1} \otimes [1\ -1]\right) \left(\mathbf{I}^{n-1} \otimes \begin{bmatrix} 2^{n-1} \\ -2^{n-1} \end{bmatrix} \right) \end{array} \right]$$

Applying the mixed product rule, $(\mathbf{A} \otimes \mathbf{B})(\mathbf{C} \otimes \mathbf{D}) = \mathbf{A}\mathbf{C} \otimes \mathbf{B}\mathbf{D}$, and simplifying we obtain

$$\mathbf{K}^n \mathbf{B}^n = \begin{bmatrix} 2\mathbf{K}^{n-1}\mathbf{B}^{n-1} & 0 \\ 0 & 2^n \mathbf{I}^{n-1} \end{bmatrix} \tag{3.15}$$

We hypothesize that $(\mathbf{K}^n)^{-1} = \frac{1}{2^n}\mathbf{B}^n$. From Equation 3.15 this is clearly true when $n = 1$. Induction on n assumes $\mathbf{K}^{n-1}\mathbf{B}^{n-1} = [2^{n-1}\mathbf{I}^{n-1}]$ substituted into Equation 3.15 yields $\mathbf{K}^n\mathbf{B}^n = [2^n\mathbf{I}^n]$. Hence $(\mathbf{K}^n)^{-1} = \frac{1}{2^n}\mathbf{B}^n$, and the theorem is proven. $\square$

Example 3.5 *For $n = 3$ the modified, normalized Haar transform inverse is*

$$[\mathbf{K}^3]^{-1} = \frac{1}{2^3} \begin{bmatrix} 1 & 1 & 2 & 0 & 4 & 0 & 0 & 0 \\ 1 & 1 & 2 & 0 & -4 & 0 & 0 & 0 \\ 1 & 1 & -2 & 0 & 0 & 4 & 0 & 0 \\ 1 & 1 & -2 & 0 & 0 & -4 & 0 & 0 \\ 1 & -1 & 0 & 2 & 0 & 0 & 4 & 0 \\ 1 & -1 & 0 & 2 & 0 & 0 & -4 & 0 \\ 1 & -1 & 0 & -2 & 0 & 0 & 0 & 4 \\ 1 & -1 & 0 & -2 & 0 & 0 & 0 & -4 \end{bmatrix} \tag{3.16}$$

As is apparent from the above example, $(\mathbf{K}^n)^{-1}$ is the transpose of $\mathbf{K}^n$ with scaling factors applied to certain columns. From the recursive structure of

Equation 3.15, one can verify that the appropriate scaling factor is 2^{n-k} where k is the $\log_2(p)$ and p is the number of non-zero entries in the column. It is clear from the definition of $\mathbf{K}^n$ that p is always a power of 2, so k is always a positive integer.

3.3 TRANSFORM PROCEDURES

The Walsh, Reed-Muller, arithmetic and Haar spectra of a Boolean function can be directly computed by appropriate matrix multiplication; however, the computational cost of this approach is generally prohibitive for functions of significant size. Fortunately, more efficient alternative techniques exist. The Cooley-Tukey type "fast" transform technique [35] is presented here. Other approaches based on DD techniques are considered in Chapter 5.

3.3.1 Fast Walsh-Hadamard Transform

The recursive definition of the Hadamard-ordered Walsh transform is the basis for a *Fast Hadamard Transform* (FHT) method analogous to a *Fast Fourier Transform* (FFT) over discrete data. Observe that

$$\mathbf{R} = \left[\begin{array}{cc} \mathbf{W}^{n-1} & \mathbf{W}^{n-1} \\ \mathbf{W}^{n-1} & -\mathbf{W}^{n-1} \end{array} \right] \left[\begin{array}{c} \mathbf{Y}^0 \\ \mathbf{Y}^1 \end{array} \right]$$

where $\mathbf{Y}^0$ and $\mathbf{Y}^1$ represent a partitioning of $\mathbf{Y}$ into two equal sized subvectors. It follows that

$$\mathbf{R} = \left[\begin{array}{c} \mathbf{W}^{n-1}\mathbf{Y}^0 + \mathbf{W}^{n-1}\mathbf{Y}^1 \\ \mathbf{W}^{n-1}\mathbf{Y}^0 - \mathbf{W}^{n-1}\mathbf{Y}^1 \end{array} \right] \tag{3.17}$$

$$= \left[\begin{array}{c} \mathbf{W}^{n-1}(\mathbf{Y}^0 + \mathbf{Y}^1) \\ \mathbf{W}^{n-1}(\mathbf{Y}^0 - \mathbf{Y}^1) \end{array} \right] \tag{3.18}$$

The above shows that the computation of the nth order transform involves the application of $(n-1)^{th}$ order transforms to two subvectors of $\mathbf{Y}$ followed by the addition and subtraction of the results. Alternatively, the transform can be computed as the addition and subtraction of two subvectors of $\mathbf{Y}$ followed by the application of two $(n-1)^{th}$ order transforms to the resultant subvectors. A similar reduction can be applied to the computation of the $(n-1)^{th}$ order transforms. Indeed the reduction can be iteratively applied down to the trivial case of applying $\mathbf{W}^0$ transforms. The result is that the objective of computing $\mathbf{W}^n\mathbf{Y}$ is reduced to a sequence of vector additions and subtractions.

At each iteration there are twice as many additions and subtractions as for the previous iteration involving subvectors of half the size. The computational work at each iteration is thus the same. In total, n iterations are involved with each havin 2^n elements yielding time complexity $O(n2^n)$. The operations can be applied in place on a single vector so the space complexity is $O(2^n)$. This is in contrast to the matrix multiplication approach where the time and space complexities are both $O(2^{2n})$.

The computational sequence arising from the above is illustrated in Figure 3.1 for the case of $n = 3$. For clarity, we show the computation as creating new vectors; however, as noted above the computation can, in practice, be performed in place. The interpretation of the *butterfly* signal flowgraphs in Figure 3.1 is as shown in Figure 3.2. In the figures a dotted line represents multiplication by -1, whereas a solid line represents a multiplicative factor of +1. Where the two lines meet at the rightmost side of the butterfly diagram, an addition operation is implied.

The FHT method represents a substantial improvement over computing the spectrum by matrix multiplication, but it is still prohibitive for large functions due to its exponential complexity. A major importance of this approach is that it forms the basis for very efficient DD-based algorithms presented in Chapter 5.

3.3.2   Fast Reed-Muller Transform

A similar approach is possible for developing a fast Reed-Muller transform since $\mathbf{M}^n$ has a recursive structure similar to that of $\mathbf{W}^n$. The situation for $n = 3$ is illustrated in Figure 3.3 with an interpretation of the signal flow subgraph shown in Figure 3.4. The computations for a fast Reed-Muller transform are over $GF(2)$.

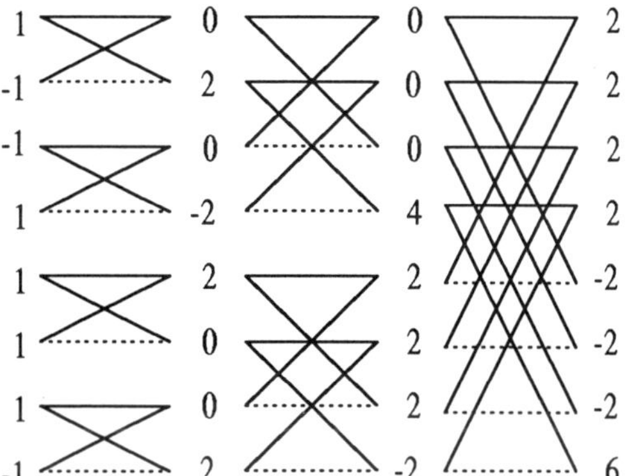

Figure 3.1 Example of the Fast Transform Computation of the Walsh Spectrum

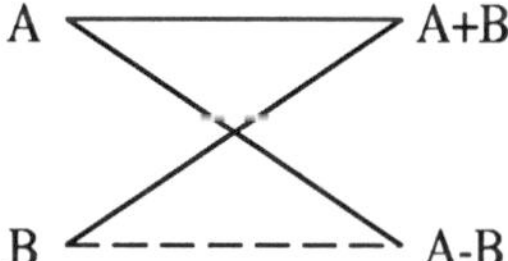

Figure 3.2 Interpretation of a "Butterfly" Signal Flowgraph for the Walsh Transform

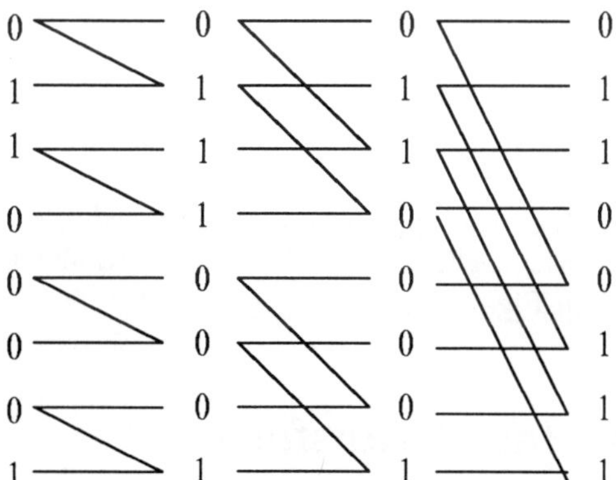

Figure 3.3 Example of the Fast Transform Computation of the Reed-Muller Spectrum

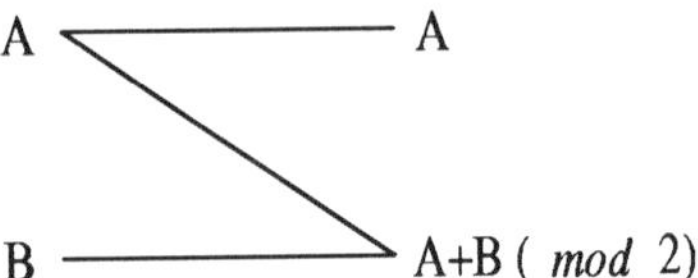

Figure 3.4 "Butterfly" Signal Flow Subgraph for Reed-Muller Transform

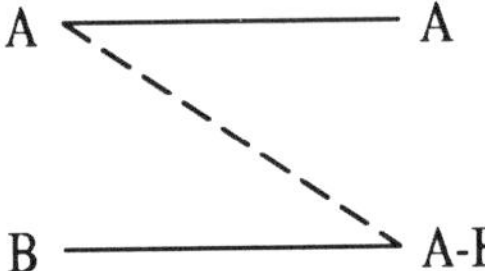

Figure 3.5 "Butterfly" Signal Flow Graph for arithmetic Transform

3.3.3 Fast Arithmetic Transform

The arithmetic transform is the integer valued version of the Reed-Muller spectral transform and can be implemented using "fast" techniques in a similar way with the difference being that each butterfly operation produces a difference of input values in the integer field instead of the sum in $GF(2)$ (*i.e.* the (*mod* 2) operation is not used). The arithmetic butterfly signal flow graph is illustrated in Figure 3.5 and an example of the "fast" arithmetic transform is shown in Figure 3.6. Note that the arithmetic transform matrix may be factored as sparse matrices in a symmetric manner; hence, Figure 3.6 contains two forms of the "fast" transform. In terms of FFTs these two formulations are analogously referred to as "decimation in time" and "decimation in frequency" respectively. The relationship between the arithmetic transform and the Reed-Muller transform is easily seen since the application of the (*mod* 2) operation on each of the arithmetic spectral coefficients in Figure 3.6 results in the Reed-Muller spectrum as shown in Figure 3.3.

3.3.4 Fast Haar Transform

The signal flowgraph for a fast normalized Haar transform can be identified directly from the recursive definition of $\mathbf{K}^n$ given in Theorem 3.13. The case for $n = 3$ is depicted in Figure 3.7. The *butterfly* structures are used as defined in the Walsh case in Figure 3.2.

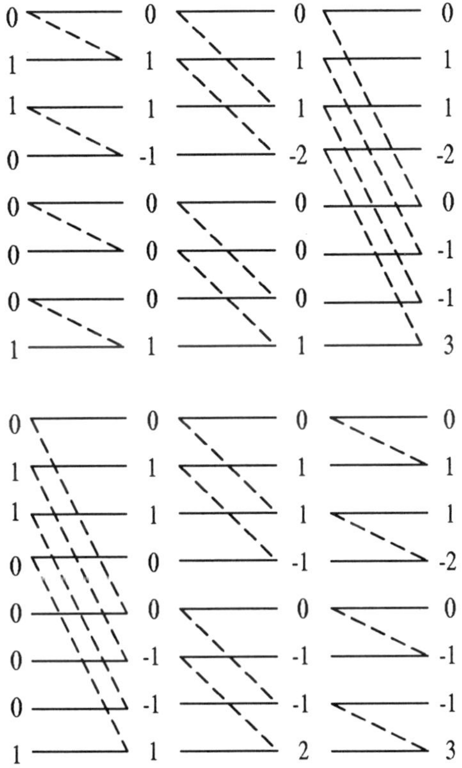

Figure 3.6 Signal Flow Graphs for the arithmetic Transform of the Example Function

Figure 3.7 depicts the normalized Haar transform. For the unnormalized transform defined by Equation 3.11, the structure is the same but appropriate multipliers must be applied in the computations.

The inverse transform has the reverse structure and, once again, appropriate multipliers must be applied in both the normalized and unnormalized cases. Figure 3.8 depicts the situation for the inverse normalized transform using the same example as Figure 3.7. In this case a value passing through a phase without going through a *butterfly* is multiplied by 2, therefore the end result is scaled by 2^3.

The Haar transform considered thus far and particularly the fast transform illustrated in Figure 3.7 is in sequency order. A drawback is that it can not be

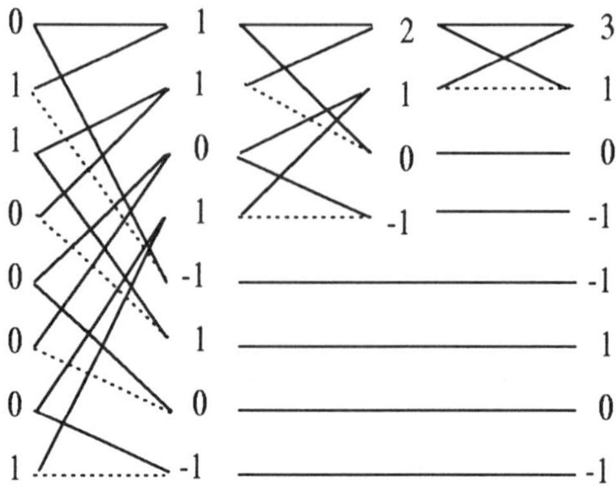

Figure 3.7 Example of the Fast Transform Computation of the Haar Spectrum

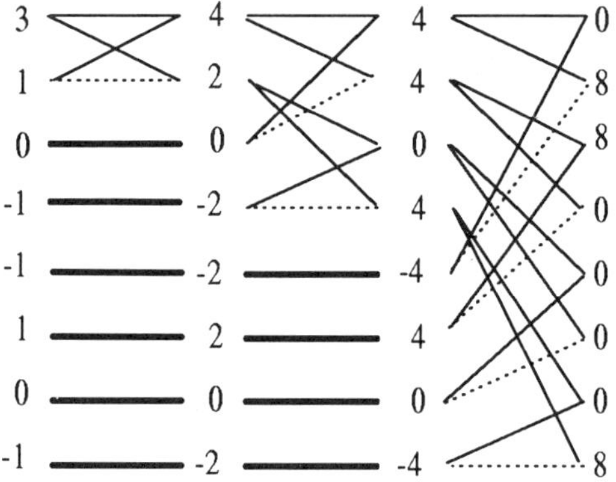

Figure 3.8 Example of the Fast Transform Computation of the Inverse Haar Transform

performed in place since, as is apparent from the flow diagram, pairs of elements are combined, and, except for the first and last element in each transform phase, the results go to other positions. An alternative is to rearrange the computations into natural order which does allow for in-place computation. It is this ordering that we employ in computing the Haar spectra using DDs in Chapter 5.

The natural order Haar transform can be defined as follows (we use $\mathbf{H}^n$ to distinguish this transform from the sequency ordered Haar transform $\mathbf{K}^n$)

$$
\begin{aligned}
\mathbf{H}^n &= \mathbf{B}^n + \mathbf{D}^n \\
\mathbf{D}^n &= \begin{bmatrix} 1 & 1 \\ 0 & 0 \end{bmatrix} \otimes \mathbf{D}^{n-1} \\
\mathbf{B}^n &= \begin{bmatrix} 1 & 0 \\ 0 & 1 \end{bmatrix} \otimes \mathbf{B}^{n-1} + \begin{bmatrix} 0 & 0 \\ 1 & -1 \end{bmatrix} \otimes \mathbf{D}^{n-1} \\
\mathbf{D}^0 &= [1], \mathbf{B}^0 = [0]
\end{aligned}
\tag{3.19}
$$

For example, for $n = 3$ the above yields

$$
\mathbf{H}^3 = \begin{bmatrix}
1 & 1 & 1 & 1 & 1 & 1 & 1 & 1 \\
1 & -1 & 0 & 0 & 0 & 0 & 0 & 0 \\
1 & 1 & -1 & -1 & 0 & 0 & 0 & 0 \\
0 & 0 & 1 & -1 & 0 & 0 & 0 & 0 \\
1 & 1 & 1 & 1 & -1 & -1 & -1 & -1 \\
0 & 0 & 0 & 0 & 1 & -1 & 0 & 0 \\
0 & 0 & 0 & 0 & 1 & 1 & -1 & -1 \\
0 & 0 & 0 & 0 & 0 & 0 & 1 & -1
\end{bmatrix}
\tag{3.20}
$$

Numbering the rows of $\mathbf{K}^3$ from 0 to 7, the rows of $\mathbf{H}^3$ adhere to the permutation [0,4,2,5,1,6,3,7]. Hence, the spectral coefficients determined using $\mathbf{H}^n$ in place of $\mathbf{K}^n$ will be similarly permuted. A more detailed analysis of this ordering is presented in [66].

Theorem 3.8 *The equations of 3.19 generate the complete set of Haar functions in natural order.*

Proof: Two initial observations for all n are that $\mathbf{D}^n$ is of order $(2^n \times 2^n)$ and consists of a top row of all 1's with 0's everywhere else and $\mathbf{B}^n$ is of order $(2^n \times 2^n)$ and has a top row of all 0's.

It is apparent from the definition of $\mathbf{H}^n$ that it can be written as

$$
\mathbf{H}^n = \begin{bmatrix}
\mathbf{B}^{n-1} + \mathbf{D}^{n-1} & \mathbf{D}^{n-1} \\
\mathbf{D}^{n-1} & \mathbf{B}^{n-1} - \mathbf{D}^{n-1}
\end{bmatrix}
\tag{3.21}
$$

It is useful to let $\mathbf{C}^n$ be a $(2^n - 1 \times 2^n)$ matrix which is $\mathbf{B}^n$ with its top row removed. $\mathbf{H}^n$ can then be written

$$\mathbf{H}^n = \begin{bmatrix} 1\,1\cdots 1 & 1\,1\cdots 1 \\ \mathbf{C}^{n-1} & \mathbf{0} \\ & \\ 1\,1\cdots 1 & -1\,-1\cdots -1 \\ \mathbf{0} & \mathbf{C}^{n-1} \end{bmatrix}$$

where $\mathbf{0}$ denotes a $(2^{n-1} - 1 \times 2^{n-1})$ matrix of 0's.

The top row of $\mathbf{H}^n$ consists of 2^n 1's and is H_0^0. The row in the middle of $\mathbf{H}^n$ is, as shown, 2^{n-1} 1's followed by 2^{n-1} -1's. It is important to note from Equation 3.14 that these are the only two Haar functions that are non-zero in both halves of the definition space. We must next show that the remaining Haar functions are also generated which we do by induction.

By definition

$$\mathbf{H}^1 = \begin{bmatrix} 1 & 1 \\ 1 & -1 \end{bmatrix}$$

Assume $\mathbf{H}^{n-1}$ includes all the $(n-1)^{th}$ order Haar functions which means $\mathbf{C}^{n-1}$ includes them all except H_0^0. This is precisely what is required since a review of Equation 3.14 shows the second and higher order Haar functions generated for n are two occurrences of the second and higher order Haar functions generated for the case of $n-1$; one in the lower half of the definition space and the second in the higher half. It follows that $\mathbf{H}^n$ includes all Haar functions. $\square$

It is also clear from the construction that $\mathbf{H}^n$ orders the Haar function in natural order, that is the function with the earliest zero-crossing first.

A fast transform technique for the naturally ordered Haar transform can be developed from the recursive structure in Equation 3.19. Figure 3.9 illustrates the situation for $n = 3$. It is interesting to observe that the structure is essentially the structure from the Walsh case with certain "butterflies" removed. The number of computations is the same as for the sequency ordered Haar transform, namely $2^n - 1$. However, a significant advantage is that the computations can be performed in place since each butterfly combines two elements and places the results in the same locations.

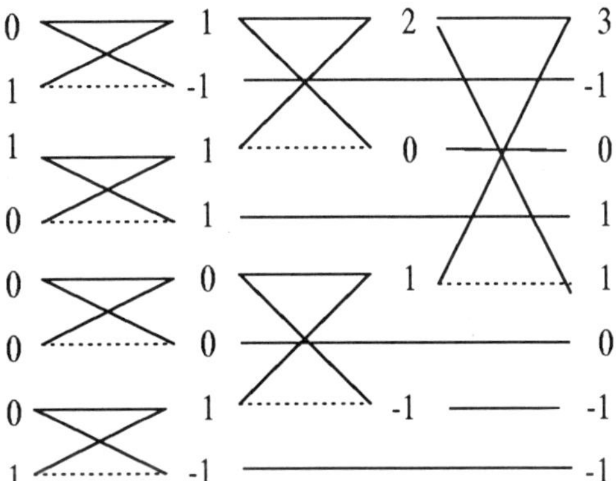

Figure 3.9 Example of the Fast Transform Computation of the Haar Spectrum in Natural Order

We next consider the inverse of $\mathbf{H}^n$, $(\mathbf{H}^n)^{-1}$.

Theorem 3.9 $(\mathbf{H}^n)^{-1} = \frac{1}{2^n}\mathbf{G}^n$ *where*

$$
\begin{aligned}
\mathbf{G}^n &= \mathbf{C}^n + \mathbf{E}^n \\
\mathbf{E}^n &= \begin{bmatrix} 1 & 0 \\ 1 & 0 \end{bmatrix} \otimes \mathbf{E}^{n-1} \\
\mathbf{C}^n &= \begin{bmatrix} 2 & 0 \\ 0 & 2 \end{bmatrix} \otimes \mathbf{C}^{n-1} + \begin{bmatrix} 0 & 1 \\ 0 & -1 \end{bmatrix} \otimes \mathbf{E}^{n-1} \\
\mathbf{E}^0 &= [1], \mathbf{C}^0 = [0]
\end{aligned}
\tag{3.22}
$$

Proof: From Equations 3.19 and 3.22 we have

$$
\begin{aligned}
\mathbf{H}^n\mathbf{G}^n &= (\mathbf{B}^n + \mathbf{D}^n)(\mathbf{C}^n + \mathbf{E}^n) \\
&= \mathbf{B}^n\mathbf{C}^n + \mathbf{B}^n\mathbf{E}^n + \mathbf{D}^n\mathbf{C}^n + \mathbf{D}^n\mathbf{E}^n
\end{aligned}
$$

From the previous theorem we know that the rows of $\mathbf{B}^n$ are Haar functions (with the exception of H_0^0) each with an equal number of +1's and -1's; thus, the sum across each row of $\mathbf{B}^n$ is 0. Hence, $\mathbf{B}^n\mathbf{E}^n = \mathbf{0}$ where $\mathbf{0}$ denotes the matrix of all 0's.

$\mathbf{C}^n$ is constructed as the transpose of $\mathbf{B}^n$ with multipliers applied to certain columns. In this way, each column of $\mathbf{C}^n$ sums to 0, so $\mathbf{D}^n\mathbf{C}^n = \mathbf{0}$ and:

$$\mathbf{H}^n\mathbf{G}^n = \mathbf{B}^n\mathbf{C}^n + \mathbf{D}^n\mathbf{E}^n$$

$\mathbf{D}^n\mathbf{E}^n$ yields a matrix with 2^n in the top left corner and 0's everywhere else.

Now,

$$\mathbf{B}^n\mathbf{C}^n = \left(\begin{bmatrix} 1 & 0 \\ 0 & 1 \end{bmatrix} \otimes \mathbf{B}^{n-1} + \begin{bmatrix} 0 & 0 \\ 1 & -1 \end{bmatrix} \otimes \mathbf{D}^{n-1}\right)$$
$$\cdot \left(\begin{bmatrix} 2 & 0 \\ 0 & 2 \end{bmatrix} \otimes \mathbf{C}^{n-1} + \begin{bmatrix} 0 & 1 \\ 0 & -1 \end{bmatrix} \otimes \mathbf{E}^{n-1}\right)$$

Simplifying, applying the Kronecker mixed product rule and multiplying the constant matrices yields:

$$\mathbf{B}^n\mathbf{C}^n = \begin{bmatrix} 2 & 0 \\ 0 & 2 \end{bmatrix} \otimes \mathbf{B}^{n-1}\mathbf{C}^{n-1} + \begin{bmatrix} 0 & 1 \\ 0 & -1 \end{bmatrix} \otimes \mathbf{B}^{n-1}\mathbf{E}^{n-1} +$$
$$\begin{bmatrix} 0 & 0 \\ 2 & -2 \end{bmatrix} \otimes \mathbf{D}^{n-1}\mathbf{C}^{n-1} + \begin{bmatrix} 0 & 0 \\ 0 & 2 \end{bmatrix} \otimes \mathbf{D}^{n-1}\mathbf{E}^{n-1}$$

As above, $\mathbf{B}^{n-1}\mathbf{E}^{n-1} = \mathbf{D}^{n-1}\mathbf{C}^{n-1} = \mathbf{0}$, therefore:

$$\mathbf{B}^n\mathbf{C}^n = \begin{bmatrix} 2 & 0 \\ 0 & 2 \end{bmatrix} \otimes \mathbf{B}^{n-1}\mathbf{C}^{n-1} + \begin{bmatrix} 0 & 0 \\ 0 & 2 \end{bmatrix} \otimes \mathbf{D}^{n-1}\mathbf{E}^{n-1}$$

We hypothesize that $\mathbf{B}^n\mathbf{C}^n$ is a diagonal matrix with a 0 in the top left entry and 2^n for every other diagonal entry. It is readily verified that this is the case for $n = 1$. Assuming that the hypothesis is true for $n - 1$ and substituting, we

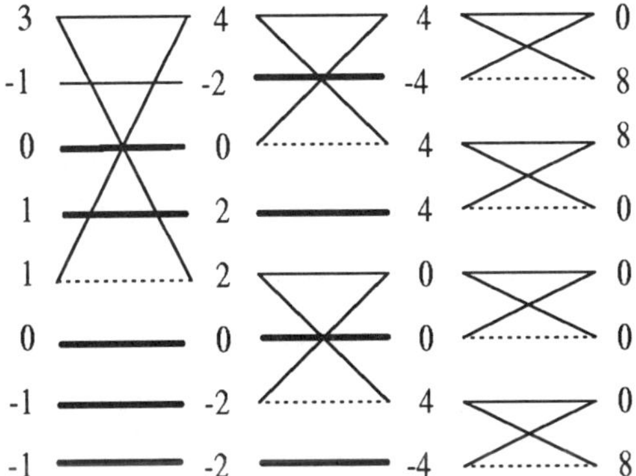

Figure 3.10 Example of the Fast Transform Computation of the Inverse Haar Transform in Natural Order

find that the hypothesis is true for all n since $\mathbf{D}^{n-1}\mathbf{E}^{n-1}$ is a matrix with 2^{n-1} in the top left corner and 0's elsewhere. Substituting this result back, we find $\mathbf{H}^n\mathbf{G}^n = 2^n\mathbf{I}^n$, and the theorem is proven. $\square$

Figure 3.10 illustrates the fast reverse transform procedure for $n = 3$. As in the sequency case, a value which passes through a phase without going through a "butterfly" must be multiplied by 2.

3.4   RELATIONSHIPS BETWEEN THE TRANSFORMS

Since each of the transforms discussed has an inverse, it is possible to create a transform from one spectral domain to another. In the worst case this is accomplished by simply passing through the Boolean functional domain. However, for some transformation types, we show direct conversion techniques.

For example, as identified in [104], if $\mathbf{S}$ is the arithmetic spectrum of a function, its Walsh spectrum $\mathbf{R}$ (in R-encoding) is given by $\mathbf{R} = \mathbf{W}^n(\mathbf{A}^n)^{-1}\mathbf{S}$. $(\mathbf{A}^n)^{-1}\mathbf{S}$ transforms the arithmetic spectrum to the functional domain after which the multiplication by $\mathbf{W}^n$ yields the Walsh spectrum. It is more efficient to treat $\mathbf{W}^n(\mathbf{A}^n)^{-1}$ as a single matrix which we can write as

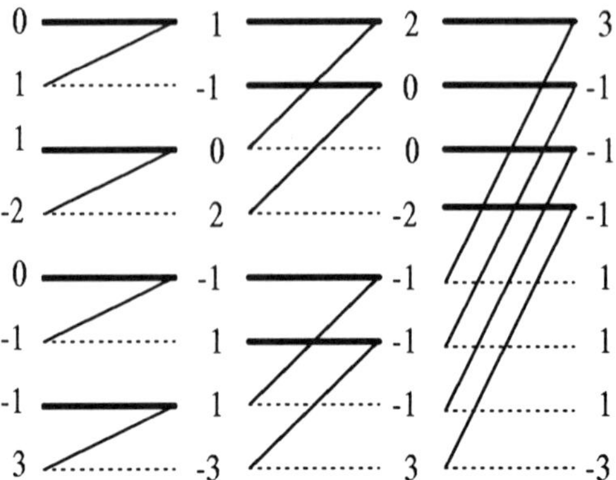

The dark lines indicate multiplication by 2.

Figure 3.11　Example of the Direct Fast Transform from the Arithmetic to the Walsh Spectrum

$$\left(\bigotimes_{i=1}^{n} \mathbf{W}^1\right)\left(\bigotimes_{i=1}^{n} \left(\mathbf{A}^1\right)^{-1}\right)$$

By the properties of the Kronecker product this can be written as

$$\bigotimes_{i=1}^{n} \mathbf{W}^1 \left(\mathbf{A}^1\right)^{-1}$$

So the transform from the arithmetic to the Walsh domain can be accomplished using the transform matrix

$$\mathbf{T}^n = \bigotimes_{i=1}^{n} \begin{bmatrix} 2 & 1 \\ 0 & -1 \end{bmatrix}$$

which can be used as the basis for a fast transform approach. This is illustrated for $n = 3$ in Figure 3.11.

Following a similar approach, it is shown that

$$\mathbf{T}^n = \frac{1}{2^n}\left(\bigotimes_{i=1}^{n}\begin{bmatrix} 1 & 1 \\ 0 & -2 \end{bmatrix}\right)$$

is a direct transform from the Walsh to the arithmetic spectral domain.

Transforming to and from the Haar domain is also possible. Next, we consider the Walsh to Haar, Haar to Walsh, arithmetic to Haar and Haar to arithmetic transforms.

Theorem 3.10 *The Walsh-Hadamard spectrum of a function can be transformed to the natural order Haar spectrum using the transform*

$$\begin{aligned}
\mathbf{T}^n &= \frac{1}{2^n}\left(\mathbf{P}^n + \mathbf{Q}^n\right) \\
\mathbf{P}^n &= \begin{bmatrix} 1 & 1 \\ 1 & -1 \end{bmatrix} \otimes \mathbf{P}^{n-1} + \begin{bmatrix} 0 & 0 \\ 0 & 2 \end{bmatrix} \otimes \mathbf{Q}^{n-1} \\
\mathbf{Q}^n &= \begin{bmatrix} 2 & 0 \\ 0 & 0 \end{bmatrix} \otimes \mathbf{Q}^{n-1} \\
\mathbf{P}^0 &= [0] \quad \mathbf{Q}^0 = [1]
\end{aligned}$$

Proof: We transform first from the Walsh to the Boolean function domain and then to the Haar domain. The transform is given by

$$\mathbf{T}^n = \mathbf{H}^n\left(\frac{1}{2^n}\mathbf{W}^n\right)$$

Employing Equation 3.19 yields

$$\mathbf{H}^n\left(\frac{1}{2^n}\mathbf{W}^n\right) = \frac{1}{2^n}\left(\mathbf{B}^n\mathbf{W}^n + \mathbf{D}^n\mathbf{W}^n\right)$$

By substitution and application of the Kronecker mixed product rule we have

$$\mathbf{B}^n\mathbf{W}^n = \begin{bmatrix} 1 & 0 \\ 0 & 1 \end{bmatrix}\begin{bmatrix} 1 & 1 \\ 1 & -1 \end{bmatrix} \otimes \mathbf{B}^{n-1}\mathbf{W}^{n-1}$$
$$+ \begin{bmatrix} 0 & 0 \\ 1 & -1 \end{bmatrix}\begin{bmatrix} 1 & 1 \\ 1 & -1 \end{bmatrix} \otimes \mathbf{D}^{n-1}\mathbf{W}^{n-1}$$

and

$$\mathbf{D}^n\mathbf{W}^n = \begin{bmatrix} 1 & 1 \\ 0 & 0 \end{bmatrix}\begin{bmatrix} 1 & 1 \\ 1 & -1 \end{bmatrix} \otimes \mathbf{D}^{n-1}\mathbf{W}^{n-1}$$

Defining $\mathbf{P}^n = \mathbf{B}^n\mathbf{W}^n$ and $\mathbf{Q}^n = \mathbf{D}^n\mathbf{W}^n$ we have

$$\mathbf{P}^n = \begin{bmatrix} 1 & 1 \\ 1 & -1 \end{bmatrix} \otimes \mathbf{P}^{n-1} + \begin{bmatrix} 0 & 0 \\ 0 & 2 \end{bmatrix} \otimes \mathbf{Q}^{n-1}$$

and

$$\mathbf{Q}^n = \begin{bmatrix} 2 & 0 \\ 0 & 0 \end{bmatrix} \otimes \mathbf{Q}^{n-1}$$

Substitution shows $\mathbf{P}^0 = [0]$ and $\mathbf{Q}^0 = [1]$, and the theorem is proven. $\qquad \square$

Theorem 3.11 *The natural order Haar spectrum of a function can be transformed to the Walsh-Hadamard spectrum using the transform*

$$\mathbf{T}^n = \frac{1}{2^n}\left(\mathbf{P}^n + \mathbf{Q}^n\right)$$
$$\mathbf{P}^n = \begin{bmatrix} 2 & 2 \\ 2 & -2 \end{bmatrix} \otimes \mathbf{P}^{n-1} + \begin{bmatrix} 0 & 0 \\ 0 & 2 \end{bmatrix} \otimes \mathbf{Q}^{n-1}$$
$$\mathbf{Q}^n = \begin{bmatrix} 2 & 0 \\ 0 & 0 \end{bmatrix} \otimes \mathbf{Q}^{n-1}$$
$$\mathbf{P}^0 = [0] \quad \mathbf{Q}^0 = [1]$$

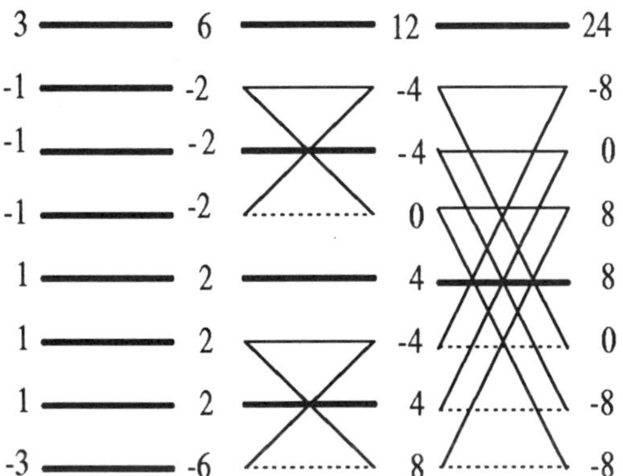

Figure 3.12 Example of the Fast Transform from the Walsh to the Haar Spectrum

Proof: The proof is analogous to the proof of Theorem 3.10.   □

These two theorems are the basis for fast transform procedures as illustrated in Figures 3.12 and 3.13. Note that while the structure of the transform is the same in each case, the butterflies in the Haar to Walsh direction are all scaled by a factor of 2. The structure is similar to the Walsh butterfly diagram presented earlier except that the first butterfly in each group is replaced by the straight through passage of the two data values scaled by 2.

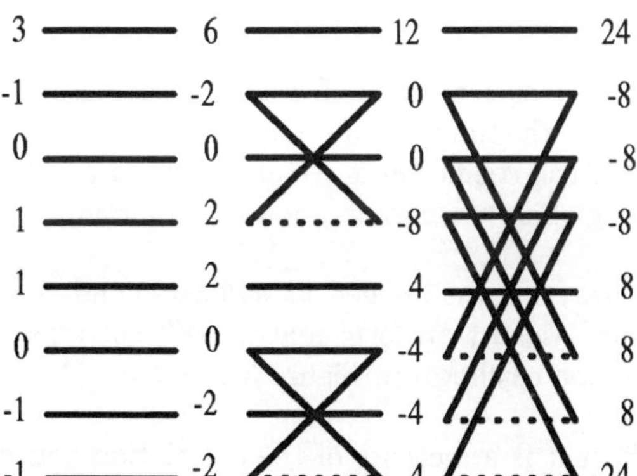

Figure 3.13 Example of the Fast Transform from the Haar to the Walsh Spectrum

This approach of combining transforms to formulate direct transformations from one spectral domain to another can not be used in the case of the Reed-Muller since those computations are carried out over $GF(2)$ while the others are carried out over the integers. However, it was shown in [104, 156] that the Reed-Muller spectral coefficients can be found by taking the (mod 2) of the absolute values of the arithmetic coefficients, a result that is not unexpected given the similar nature of the two transform matrices.

3.5　SPECTRAL PROPERTIES

In this section a qualitative interpretation of spectral coefficients is given. Although these interpretations are not mathematically rigorous, we believe that this viewpoint may lead to an intuitive feeling for the use of various spectra in CAD applications.

3.5.1　Interpretation of Spectral Coefficients

Since each of the spectral transforms considered previously has an inverse, it is clear that the information content of the Boolean domain specification for a function and the spectra computed from it have equivalent overall information content. However, the information is organized differently so that each of the spectrum types illuminates information that differs from the others and also from the Boolean logic domain specification.

In the Boolean domain each element of $\mathbf{Y}$ or $\mathbf{Z}$ provides the function value at a specific assignment of the function arguments. This is purely *local* information. In contrast, each Walsh coefficient is computed using all of the function values and thus provides global information about the function.

In particular, interpreting the integer $+1$ as logic-0 and -1 as logic-1, each row of the Rademacher-Walsh transform matrix $\mathbf{W}^n$ corresponds to a particular exclusive-OR function as shown in Figure 3.14 for $n = 3$.

Each Walsh coefficient is a measure of the correlation between the given logic function and the exclusive-OR function corresponding to the spectral coefficient. For the case of S-encoding, the coefficient value ranges from $+2^n$ to -2^n where a positive value represents correlation to the exclusive-OR function and a negative value represents correlation to the inverse function (equivalence or

$$\begin{bmatrix} 1 & 1 & 1 & 1 & 1 & 1 & 1 & 1 \\ 1 & -1 & 1 & -1 & 1 & -1 & 1 & -1 \\ 1 & 1 & -1 & -1 & 1 & 1 & -1 & -1 \\ 1 & -1 & -1 & 1 & 1 & -1 & -1 & 1 \\ 1 & 1 & 1 & 1 & -1 & -1 & -1 & -1 \\ 1 & -1 & 1 & -1 & -1 & 1 & -1 & 1 \\ 1 & 1 & -1 & -1 & -1 & -1 & 1 & 1 \\ 1 & -1 & -1 & 1 & -1 & 1 & 1 & -1 \end{bmatrix} \begin{matrix} [0] \\ [x_1] \\ [x_2] \\ [x_1 \oplus x_2] \\ [x_3] \\ [x_1 \oplus x_3] \\ [x_2 \oplus x_3] \\ [x_1 \oplus x_2 \oplus x_3] \end{matrix}$$

Figure 3.14 Example of Rademacher-Walsh Transformation Matrix for $n = 3$

exclusive-NOR). Based on this interpretation, it is not surprising that, while traditional Boolean logic techniques focus on AND, OR and NOT operations, logic design techniques employing the Walsh spectra can illuminate exclusive-OR structures more directly. The **R** spectrum is a linear translation of the **S** spectrum and thus contains identical information content, albeit in a different form.

The Reed-Muller spectrum contains certain coefficients that depend on all the function values, whereas others only depend on a particular structured subset. Thus, they convey a blend of global and localized information.

Like the Reed-Muller spectrum, the Haar spectral coefficients provide a mix of global and localized information. Given $\mathbf{S} = \mathbf{K}^n \mathbf{Z}$, the Haar spectrum of a function $f(X)$, the resulting spectral coefficients have the following interpretation

1. s_0: correlation to the constant 0 function;

2. s_1: correlation with input function x_1, range is $\pm 2^n$;

3. s_2: correlation with input function x_2 within the restricted space $\overline{x}_2$, range is $\pm 2^{n-1}$;

4. s_3: correlation with input function x_2 within the restricted space x_2, range is $\pm 2^{n-1}$;

5. s_4: correlation with input function x_3 within the restricted space $\overline{x}_1\overline{x}_2$, range is $\pm 2^{n-2}$;

$$
\begin{bmatrix}
1 & 1 & 1 & 1 & 1 & 1 & 1 & 1 \\
1 & -1 & 1 & -1 & 1 & -1 & 1 & -1 \\
1 & 1 & -1 & -1 & 0 & 0 & 0 & 0 \\
0 & 0 & 0 & 0 & 1 & 1 & -1 & -1 \\
1 & -1 & 0 & 0 & 0 & 0 & 0 & 0 \\
0 & 0 & 1 & -1 & 0 & 0 & 0 & 0 \\
0 & 0 & 0 & 0 & 1 & -1 & 0 & 0 \\
0 & 0 & 0 & 0 & 0 & 0 & 1 & -1
\end{bmatrix}
\quad
\begin{matrix}
[f] \\
[x_1] \\
[x_2 \cdot f_{\overline{x}_1}] \\
[x_2 \cdot f_{x_1}] \\
[x_3 \cdot f_{\overline{x}_1 \overline{x}_2}] \\
[x_3 \cdot f_{\overline{x}_1 x_2}] \\
[x_3 \cdot f_{x_1 \overline{x}_2}] \\
[x_3 \cdot f_{x_1 x_2}]
\end{matrix}
$$

Figure 3.15 Example of Normalized Haar Transformation Matrix for $n = 3$

6. s_5: correlation with input function x_3 within the restricted space $x_1 \overline{x}_2$, range is $\pm 2^{n-2}$;

$\vdots$

7. s_{2^n-1}: correlation with input function x_3 within the restricted space $x_1 x_2$, range is ± 2.

As an example of the normalized Haar transform, consider the transformation matrix for a 3-variable function as is shown in Figure 3.15. Each of the constituent Boolean functions are given to the left of their respective row vectors. Note that the matrix contains a 0 value in addition to the 1 and -1 quantities which represent logic levels (assuming S-encoding is used). Since some of the rows represent constituent functions that are cofactors, the output space is less than 2^3 and the presence of a 0 value acts as the null-space of the respective functions shown to the left in Figure 3.15.

3.5.2   Spectral Translation

Since the Walsh functions are interpreted as an encoding of the set of exclusive-OR functions and the Walsh spectral coefficients measure the correlation of the function transformed to those exclusive-OR functions, it is not surprising that Walsh spectral techniques illuminate and deal with the exclusive-OR far better than do traditional logic design methods. One notable example is *spectral translation* [90, 92]. The spectral translation operations show how certain operations in the Boolean domain are reflected in the Walsh spectral domain. Details are given in Table 3.1. The first three are common operations in the Boolean domain and are the basis of the well-known *NPN* classification of

Table 3.1 Spectral Translation Operations

Translation	Functional	S-encoding	R-encoding
(i) input permutation	$x_i \leftrightarrow x_j$	$s_{i\alpha} \leftrightarrow s_{j\alpha}$	$r_{i\alpha} \leftrightarrow r_{j\alpha}$
(ii) input negation	$x_i \leftarrow \overline{x_i}$	$s_{i\beta} \leftarrow -s_{i\beta}$	$r_{i\beta} \leftarrow -r_{i\beta}$
(iii) function negation	$f \leftarrow \overline{f}$	$s_0 \leftarrow -s_0$ $s_\gamma \leftarrow -s_\gamma$	$r_0 \leftarrow 2^n - r_0$ $r_\gamma \leftarrow -r_\gamma$
(iv) input translation	$x_i \leftarrow x_i \oplus x_j$	$s_{i\alpha} \leftrightarrow s_{ij\alpha}$	$r_{i\alpha} \leftrightarrow r_{ij\alpha}$
(v) output translation	$f \leftarrow f \oplus x_i$	$s_{i\beta} \leftrightarrow s_\beta$	$r_0 \leftarrow 2^{n-1} - r_i$ $r_i \leftarrow r_0 - 2^{n-1}$ $r_{i\beta} \leftrightarrow r_\beta$

α denotes a subset of $\{1, 2, \ldots, n\}$ which does not contain i or j.

β denotes a subset of $\{1, 2, \ldots, n\}$ which does not contain i.

γ denotes a subset of $\{1, 2, \ldots, n\}$ other than the empty set.

Each assignment or interchange is performed for every applicable set of subscripts.

functions [92]. The latter two employ the exclusive-OR operation. Translation (iv) is of particular interest since it is the basis for a decision diagram variable reordering technique referred to as *linear transformation* or *linear sifting* [115] which will be discussed in Chapter 6. In that work, the translation is referred to as *linear translation* reflecting the fact that the exclusive-OR is a linear operator.

3.5.3 Other Properties

Several other properties of Boolean functions can be analyzed in the spectral domain. Function symmetry, for example, is extensively studied from the point of view of the Walsh spectral properties in [92]. The work in [92] also considers the use of the Walsh spectral properties in function classification, logic synthesis and fault detection.

3.6 SUMMARY

This chapter has introduced the basic background concerning the spectra of logic functions with particular emphasis on the Walsh, the Reed-Muller, the arithmetic and the Haar spectra. The basic definitions have been presented as well as matrix and fast transform techniques for the computation of these spectra. Transformation techniques have also been explored between various spectral domains. Subsequent chapters present DD methods for the computation of the various spectra and examine their application in spectral based CAD methods.

The presentation here and the development throughout the rest of this book are directed toward binary systems. However, spectral techniques and DDs can be applied to multiple-valued logic functions (*i.e.* a function whose inputs and whose output assume p distinct values with $p \geq 2$).

4

DECISION DIAGRAMS

This chapter focuses on the use of various forms of *Decision Diagrams* (DDs) for the representation of the spectra of Boolean functions. Several DD variants will be described with their unique characteristics discussed followed by a description of their usage in representing Boolean functions in the binary as well as the spectral domains.

4.1   DECISION DIAGRAM DEFINITIONS

First, a purely structural definition of a decision diagram DD over a set of Boolean variables is given. The semantics are particular to the kind of diagram and will be described as the different diagram types are introduced throughout the chapter.

A DD over a set of Boolean variables X and a nonempty finite set of terminal values T is a connected, acyclic directed graph $G = (V, E)$ with exactly one root vertex and the following properties:

- Each vertex in G is either a terminal or a nonterminal vertex.

- Each nonterminal vertex v is labeled with a variable from X called the *index* of v (denoted *index*(v)) and has two outgoing edges leading to the two successors (not necessarily distinct) of v which are denoted as *low*(v) and *high*(v).

- Each terminal vertex v is labeled by a (*value*(v)) from T (denoted *value*) and has no outgoing edges, hence no successors.

51

The edge from a nonterminal v to $low(v)$ is denoted $e_{low}(v)$ and the edge from a nonterminal v to $high(v)$ is denoted $e_{high}(v)$. In diagrams, low (high) edges are labeled 0 (1). When edge markings are not present, the leftmost edge is $low(v)$ and the rightmost edge is $high(v)$.

The following properties are applicable to all the DDs considered here:

- The DD given by G is *complete* if each variable in X is encountered exactly once on each path from the root to a terminal in G.

- The DD given by G is *free* if each variable in X appears at most once on each path from the root to a terminal in G.

- The DD given by G is *ordered* if it is free and the variables in X on each path from the root to a terminal in G adhere to a fixed ordering.

  More precisely, the variable ordering of an ordered DD is a mapping $\pi : \{1,\ldots,n\} \rightarrow X$ where $\pi(i)$ denotes the ith variable in the order. It follows that for any nonterminal vertex v, $index(low(v)) = \pi(j)$ with $j > \pi^{-1}(index(v))$ for $low(v)$ a nonterminal and $index(high(v)) = \pi(k)$ with $k > \pi^{-1}(index(v))$ for $high(v)$ a nonterminal (j and k may be equal but need not be). It also follows that if two variables appear on two paths from the root to nonterminal vertices, they occur in the same order.

- The DD represented by G is *reduced* if (i) no nonterminal vertex v has $low(v) = high(v)$ and (ii) no two nonterminal vertices v_i and v_j satisfy:

$$index(v_i) = index(v_j),\ low(v_i) = low(v_j),\ \text{and}\ high(v_i) = high(v_j)$$

Throughout this work we are mostly concerned with *Reduced Ordered DDs* (RODD). If the variable ordering is not specified, it is assumed to be the natural ordering $\{1, 2, \ldots, n\}$.

4.2   DECISION DIAGRAM VARIETIES

Several types of DDs have been proposed in the area of CAD, among them being bit-level DDs such as BDDs [15], FDDs [98] and KFDDs [51]. While the aforementioned types of DDs are suitable for representing Boolean functions at the bit-level and have proven useful in many applications in CAD, recently DDs that represent integer-valued functions (*i.e.*, word level DDs) such as MTBDDs

[34] (also referred to as ADDs [4]), BMDs [19], KBMDs [41] and HDDs [33] (note that HDDs are a superset of KBMDs) are attracting more interest since they can offer more compact representations for some functions. Also, the use of DD edge annotations can also reduce the size. Commonly used DDs with edge annotations include EVBDDs [106], FEVBDDs [151], *BMDs [19] and K*BMDs [45].

These various forms of DDs are classified as either bit-level or word-level DDs with or without edge annotations. Bit-level DDs have terminal values corresponding to binary constants while word-level DDs have terminal vertices annotated with integers or some other non-binary value. Both types of DDs are considered here with examples given. For a more detailed overview see [41].

4.2.1 Decompositions and Spectral Interpretation

Typically, DDs are defined based on the following characteristics:

i) vertex annotations

ii) edge annotations

iii) vertex decomposition rules

iv) graph reduction rules

In practice DDs are synthesized directly from a netlist or some other relatively compact representation of a discrete function. For ease of understanding, we will assume that DDs are derived by the application of the reduction rules defined above to *Decision Trees* (DTs). An example of a DT where the nonterminal vertices obey the Shannon decomposition rule is shown in Figure 4.1. Notice that the terminal values are all possible evaluations of the function represented in DT form.

In a spectral interpretation of DDs, DTs can be defined in terms of linear transformation operations at each nonterminal vertex [150]. Specifically, linear transformation matrices that can be defined in a recursive manner (a Kronecker product based definition) map well to the structure of a DT, which in turn can be used to derive a DD. This is convenient since most commonly used spectral transforms can be derived using such matrices.

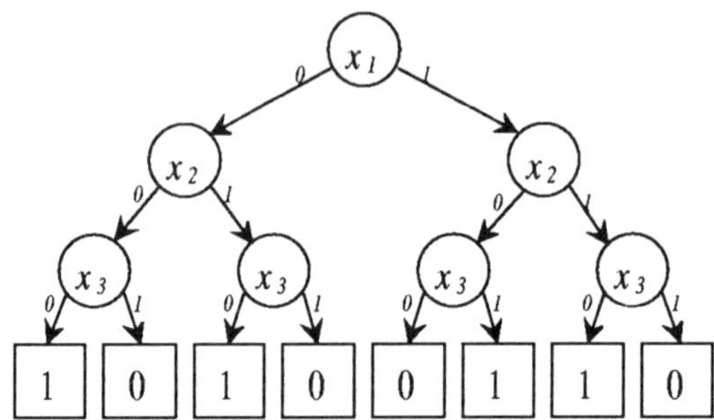

Figure 4.1   Example Shannon Tree

The recursive structure of a transform matrix based on Kronecker products [74] permits the spectral representation of the decomposition of f in terms of a DT. Each 2×2 submatrix $\mathbf{A^1}$ of such a spectral transformation matrix represents a decomposition rule performed at the nonterminal vertices at each level of a DT. When such a DT is reduced to a DD, the resulting structure may be referred to as a *Spectral Decision Diagram* (SDD). Various DDs reported in the literature can be described by the $\mathbf{A^1}$ matrix used to reduce DTs into DDs. The edge-valued versions of SDDs may be interpreted through corresponding partial spectral transforms [149]. With this viewpoint the remainder of this chapter will survey various common types of DDs classifying them according to the decomposition type used. This will allow the correspondence between the spectrum and the DD representing a function to become clear.

4.3   BIT-LEVEL DECOMPOSITIONS

In terms of bit-level DDs, all arithmetic operations are performed over $GF(2)$ (addition is the binary exclusive-OR, and multiplication is the binary AND operation). There exist six 2×2 possible transformation matrices labeled (A) through (F) as shown in Figure 4.2.

The three matrices labeled (A) through (C) in the top of Figure 4.2 correspond to the Shannon, *positive Davio* (pD) and *negative Davio* (nD) decompositions, respectively. As is further elaborated in the following chapter, the pD and nD are also known as the *positive polarity Reed-Muller* and *negative polarity Reed-Muller* transforms. Each matrix on the bottom row of Figure 4.2 ((D) through (F)) represents a decomposition that simultaneously performs a transformation while also inverting the polarity of a function variable.

$$(A) \begin{bmatrix} 1 & 0 \\ 0 & 1 \end{bmatrix} \quad (B) \begin{bmatrix} 1 & 0 \\ 1 & 1 \end{bmatrix} \quad (C) \begin{bmatrix} 1 & 1 \\ 0 & 1 \end{bmatrix}$$

$$(D) \begin{bmatrix} 0 & 1 \\ 1 & 0 \end{bmatrix} \quad (E) \begin{bmatrix} 0 & 1 \\ 1 & 1 \end{bmatrix} \quad (F) \begin{bmatrix} 1 & 1 \\ 1 & 0 \end{bmatrix}$$

Figure 4.2 Possible 2 × 2 Bit-level Transformation Matrices

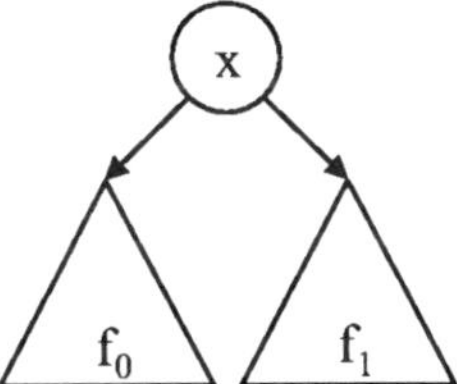

Figure 4.3 Shannon Decomposition of a DD Vertex

4.3.1 Binary Decision Diagrams

A BDD can be interpreted in the spectral domain as obeying the decomposition rule given by matrix (A) in Figure 4.2. This decomposition matrix, when multiplied with a column vector containing the Shannon cofactor functions about some variable x, yields the functions represented by the subtrees $low(x)$ and $high(x)$. Since matrix (A) is actually the 2×2 identity matrix $\mathbf{I}$, the subtree $low(x)$ represents the cofactor function f_0 while the subtree $high(x)$ represents f_1. Equation 4.1 illustrates this type of decomposition mathematically, and Figure 4.3 shows a graphical interpretation.

$$\begin{bmatrix} 1 & 0 \\ 0 & 1 \end{bmatrix} \begin{bmatrix} f_0 \\ f_1 \end{bmatrix} = \begin{bmatrix} f_0 \\ f_1 \end{bmatrix} \tag{4.1}$$

4.3.2 Functional Decision Diagrams

Functional Decision Diagrams (FDDs) were proposed in [98] as an alternative representation for Boolean functions which sometimes can be smaller than BDDs. An FDD utilizes the pD decomposition described by matrix (B) in

Figure 4.2. For BDDs and FDDs $low(x) = f_0$; however, the difference between these two representations is that $high(x) = f_0 \oplus f_1$. Equation 4.2 illustrates this type of decomposition mathematically, and Figure 4.4 shows a graphical interpretation for the pD decomposition (the nD decomposition is also illustrated on the right). FDDs that include nD decomposition types were introduced in [52].

$$\begin{bmatrix} 1 & 0 \\ 1 & 1 \end{bmatrix} \begin{bmatrix} f_0 \\ f_1 \end{bmatrix} = \begin{bmatrix} f_0 \\ f_0 \oplus f_1 \end{bmatrix} \tag{4.2}$$

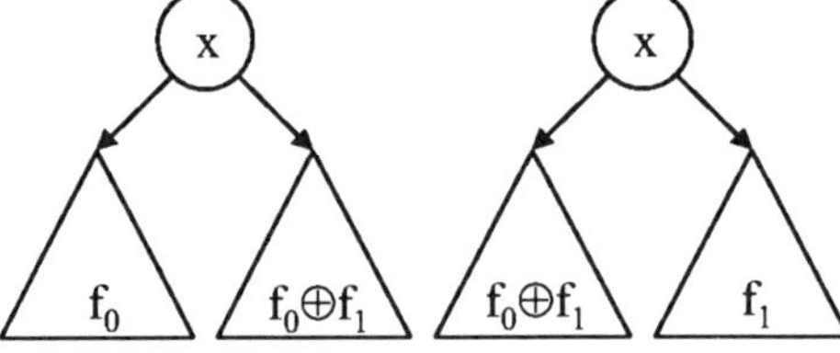

Figure 4.4   Positive and Negative Davio Decompositions of a DD Vertex

Since the pD decomposition matrix ((B) in Figure 4.2) is identical to the positive-polarity Reed-Muller transformation [123], the FDD is also the SDD for this transformation. The FDD can be interpreted as a graphical representation of the positive polarity Reed-Muller spectrum for a binary function.

4.3.3   Kronecker Functional Decision Diagrams

The *Kronecker Functional Decision Diagram* (KFDD) combines the three types of bit-level decompositions represented by matrices (A), (B) and (C) in Figure 4.2 [51] (these are the Shannon, positive Davio and negative Davio decompositions). KFDDs are ordered and reduced and require that a particular variable utilize the same decomposition everywhere it occurs in the graph. These restrictions allow KFDDs to remain canonical for a particular function, variable ordering and *Decomposition Type List* (DTL). The DTL specifies which of the three possible decompositions are assigned to each variable of the function being represented by the KFDD. KFDD size reduction requires the determination of a DTL as well as an ordering of the variables.

$$(A) \begin{bmatrix} 1 & 1 \\ 1 & -1 \end{bmatrix} \quad (B) \begin{bmatrix} 1 & 0 \\ -1 & 1 \end{bmatrix} \quad (C) \begin{bmatrix} 0 & 1 \\ 1 & -1 \end{bmatrix}$$

$$(D) \begin{bmatrix} 1 & 0 \\ 0 & 1 \end{bmatrix} \quad (E) \begin{bmatrix} 1 & 0 \\ 1 & 1 \end{bmatrix} \quad (F) \begin{bmatrix} 0 & 1 \\ 1 & 1 \end{bmatrix}$$

Figure 4.5   Possible 2×2 Word-level Transformation Matrices

Extensions of the KFDD have been proposed and studied. The *Pseudo-Kronecker Functional Decision Diagram* (PKFDD) drops the restriction that all common variables must utilize the same decomposition type [140]. PKFDDs can be smaller than KFDDs; however, in general, it is very difficult to find a minimal PKFDD. Another closely related DD is the *Ternary Decision Diagram* (TDD) [137]. TDDs are similar to KFDDs, except that each nonterminal vertex has three outgoing edges pointing to subtrees that represent f_0, f_1 and $f_0 \oplus f_1$. TDDs are canonical however they contain redundant information about the function being represented.

4.4   WORD-LEVEL DECOMPOSITIONS

Word-level DDs may be obtained by invoking the reduction rules on a DT with word-valued terminal vertices. For the case of binary-valued functions, a corresponding word-valued DT may be obtained by using S or R encoding. The particular type of word-level DD obtained depends on the decomposition type at each nonterminal vertex. As with the bit-level cases, the decomposition types can be described succinctly with 2×2 transformation matrices. Although all possible non-singular matrices of size 2×2 over $GF(2)$ were considered in the bit-level case, an infinite number exist over the integer field. In the area of spectral methods for VLSI CAD, practitioners typically use normalized transformation matrices that are composed of elements with the value $\{-1, 0, 1\}$. Restricting the matrix element values to $\{-1, 0, 1\}$ leads to 81 possible matrices of interest [63]. Of the 81 matrices, six are shown in Figure 4.5 as all of the other matrices are either singular or lead to DDs that are isomorphic to those produced by one of these six [30].

DDs resulting from the use of matrix (A) are SDDs corresponding to the Walsh or the Haar spectral transformations. The difference between these two transformations is that an equal number of transformations are performed at each level of the DT in the case of the Walsh, while a varying number are performed for the Haar. There are three other 2×2 matrices that could be obtained by permuting the matrix (A). These are termed *multi-polarity generalized Walsh transformations* in [64] and will not be considered further here.

The decomposition represented by matrix (B) in Figure 4.5 corresponds to the *arithmetic transform* [85] and is also referred to as the *inverse integer Reed-Muller transform* [33] and the integer-valued positive Davio (pD) [43]. The transform matrix obtained from the transpose of matrix (B) is the *probability transform* [104]. The arithmetic transform has the useful property of describing binary-valued functions as pseudo-Boolean functions and can also be used to determine the positive polarity Reed-Muller SDD by taking the $(mod\ 2)$ values of the terminal vertex annotations.

Matrix (C) in Figure 4.5 represents the integer-valued negative Davio (nD) decomposition. Matrices (B) and (C) in Figure 4.5 are the integer-valued analogs to the bit-level decomposition matrices (B) and (C) in Figure 4.2.

Matrix (D) in Figure 4.5 is the two-dimensional identity matrix I and represents a Shannon decomposition for a word-level DD.

The decompositions represented by matrices (E) and (F) are the positive polarity integer-valued Reed-Muller decomposition and the negative polarity integer-valued Reed-Muller decomposition, respectively. Matrix (E) also happens to be the inverse of the arithmetic decomposition matrix (B) and has also been identified as the *adding transform* in [63].

4.4.1   Multi-terminal Binary Decision Diagrams

Multi-Terminal Binary DDs (MTBDDs) [29] are the word-level analogs to the bit-level BDDs represented by Shannon decompositions (matrix (D) in Figure 4.5) at each nonterminal vertex. MTBDDs are also referred to as *Algebraic Decision Diagrams* (ADDs) [4] since they were independently developed by another research group. In general, MTBDDs are useful for representing binary-valued variable, integer-valued functions. They have been applied to problems such as the computation of Markov chain steady state probabilities [81] and as representations for multi-output Boolean functions by grouping all outputs together and representing them as an integer. Of particular interest to read-

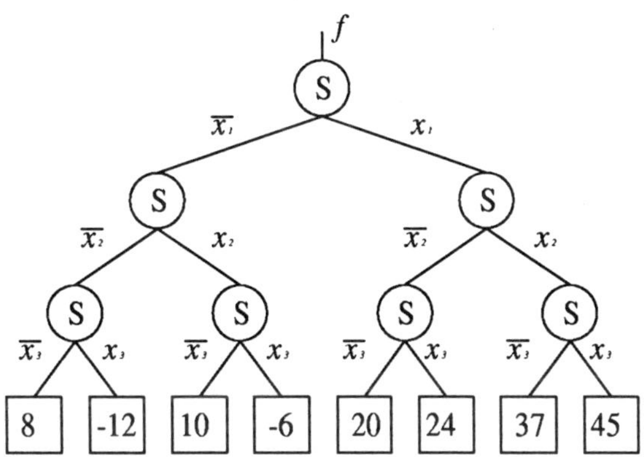

Figure 4.6   Example MTBDD for Function f

ers of this book, MTBDDs may be used to represent integer-valued matrices and vectors allowing linear algebra manipulations to be implemented as DD algorithms.

Figure 4.6 shows a diagram representing a MTBDD for the three-variable function, $f(x_1, x_2, x_3)$, $x_i \in \{0, 1\}$, given by Equation 4.3. Note that this example originally appeared in [18].

$$
\begin{aligned}
f &= 8\bar{x}_1\bar{x}_2\bar{x}_3 - 12\bar{x}_1\bar{x}_2 x_3 + 10\bar{x}_1 x_2\bar{x}_3 - 6\bar{x}_1 x_2 x_3 \\
&+ 20 x_1\bar{x}_2\bar{x}_3 + 24 x_1\bar{x}_2 x_3 + 37 x_1 x_2\bar{x}_3 + 45 x_1 x_2 x_3
\end{aligned}
\tag{4.3}
$$

The nonterminal vertices of MTBDDs obey the Shannon decomposition rule denoted by the 2×2 identity matrix I, thus they are labeled with an "S". This results in terminal vertex values that are coefficients of all the possible minterms of the binary variables as is seen in Equation 4.3. From a spectral point of view, the MTBDD is a graphical representation of a function decomposed by basis functions that are all possible minterms. The MTBDD is the word-level counterpart of the BDD and reduction techniques for MTBDDs are identical to those used for BDDs since both utilize the Shannon decomposition.

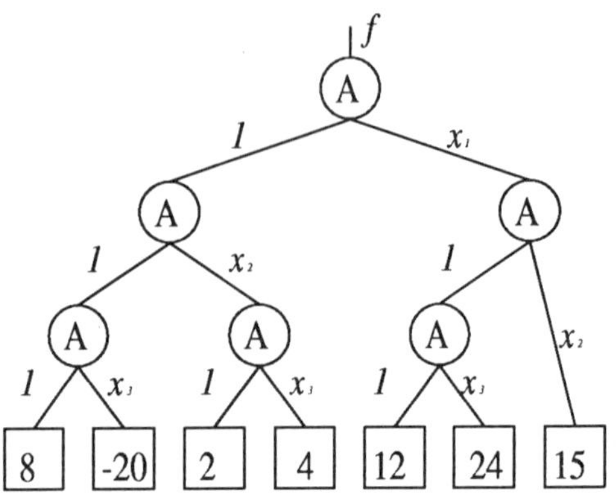

Figure 4.7   Example BMD for Function f

4.4.2   Binary Moment Diagrams

Binary Moment Diagrams (BMDs) [18, 19] utilize decomposition relations denoted by matrix (B) in Figure 4.5. Therefore, a BMD is an SDD obeying the arithmetic transform. The terminal vertex values of a BMD are the arithmetic spectral coefficients [150]. Since the arithmetic transform is the integer-valued positive Davio decomposition (or integer-valued, positive polarity Reed-Muller transform), reduction of these DDs may be carried out analogously to the bit-level FDDs [51]. From a spectral point of view, the terminal values can be interpreted as the coefficients of the function decomposed about the positive polarity Reed-Muller basis functions which consist of all possible unique positive unate cubes (2^n in total). Since the positive polarity Reed-Muller basis functions for a function of one variable are 1 and x, the function represented for a single vertex BMD is given by Equation 4.4.

$$f = \begin{bmatrix} 1 & 0 \\ -1 & 1 \end{bmatrix} \begin{bmatrix} f_0 \\ f_1 \end{bmatrix} \begin{bmatrix} 1 & x \end{bmatrix} = 1 \cdot f_0 + x \cdot (f_1 - f_0) \tag{4.4}$$

Figure 4.7 shows a diagram representing the BMD for the same function f as was used in Figure 4.6. The nonterminal vertices are labeled with an "A" to denote that the decomposition rule at each node corresponds to the 2×2 arithmetic transform. By interpreting the logical complement of each literal in Equation 4.3 as a numerical complement (replacing all $\overline{x}_i$ with $(1 - x_i)$), a pseudo-Boolean expression is obtained. When the pseudo-Boolean expression is simplified, Equation 4.5 results.

$$f = 8 - 20x_3 + 2x_2 + 4x_2x_3 + 12x_1 + 24x_1x_3 + 15x_1x_2 \qquad (4.5)$$

As is described in [18], the resulting zero-valued constant node is not shown in Figure 4.7 since it corresponds to a zero-valued arithmetic transform coefficient and is not needed to evaluate f. The coefficients of all the cubes in Equation 4.5 are the arithmetic spectral values (with zero-valued coefficients omitted). The zero-valued arithmetic spectral coefficient corresponds to the Reed-Muller basis function $x_1x_2x_3$ which is not present in Equation 4.5. Due to this phenomenon, the arithmetic transform based SDD can allow for a reduction in size since the BMD as compared to the MTBDD allows for the removal of a nonterminal and a terminal vertex in the example. The BMD representing a pseudo-Boolean function can easily be transformed into a bit-level FDD representing its' Boolean function counterpart by taking the (mod 2) value of each terminal vertex and reducing the resulting DD thus yielding a graphical representation of the positive polarity Reed-Muller spectrum.

4.4.3   Kronecker Binary Moment Diagrams

Kronecker Binary Moment Diagrams (KBMDs) [45, 46] are the word-level counterparts of bit-level KFDDs for binary functions [51]. For KBMD representations, the decomposition of f is performed by freely choosing any of the three matrices (B), (C) and (D) in Figure 4.5, while enforcing the constraint that all vertices labeled with the same variable utilize the same decomposition. The corresponding DT for a KBMD is one in which each level of the tree corresponds to one of the three possible decomposition types. From this point of view, given a variable ordering, a DTL may be used to uniquely describe a particular instance of a KBMD for a function. Reduction rules are analogous to those used for bit-level KFDDs.

For MTBDD and BMD representations, a DD with a specified variable order uniquely describes any function f. The advantage in using a KBMD is given by increasing the number of possible different DDs that can result through the use of varying DTLs and variable orders for f. This leads to 3^n different possible DTLs and, when variable orderings are considered, $n!3^n$ different KBMDs. It has been shown that KBMDs can represent many functions in a compact

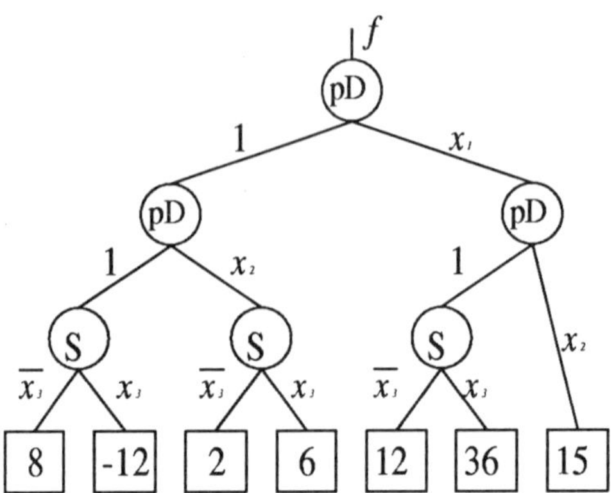

Figure 4.8   Example KBMD for Function f

manner [45]; however, finding the optimal DTL and an associated variable ordering can be a difficult problem to solve.

Figure 4.8 shows a KBMD representing the example function f with a variable ordering of $x_1 \prec x_2 \prec x_3$ and a corresponding DTL of pD $\prec$ pD $\prec$ S. Since a KBMD may utilize pD, nD or S decompositions, the non-terminal vertices are labeled with the particular decomposition that they obey. The pD decomposition for integer-valued functions is identical to the arithmetic decomposition thus, the nodes labeled with "pD" could also be labeled with "A" instead. Equation 4.6 is the example function in the form that is obtained directly from the KBMD.

$$f = 8\overline{x}_3 - 12x_3 + 2x_2\overline{x}_3 + 6x_2x_3 + 12x_1\overline{x}_3 + 36x_1x_3 + 15x_1x_2 \tag{4.6}$$

A spectral interpretation of the example KBMD is that it represents an SDD whose 8×8 transformation matrix is computed as shown in Equation 4.7.

$$\begin{bmatrix} 1 & 0 \\ -1 & 1 \end{bmatrix} \otimes \begin{bmatrix} 1 & 0 \\ -1 & 1 \end{bmatrix} \otimes \begin{bmatrix} 1 & 0 \\ 0 & 1 \end{bmatrix} \tag{4.7}$$

The basis functions φ_i corresponding to the transformation matrix are shown in the following set of equations.

$$\varphi_0 = \overline{x}_3, \quad \varphi_1 = x_3, \quad \varphi_2 = x_2\overline{x}_3, \quad \varphi_3 = x_2 x_3,$$
$$\varphi_4 = x_1\overline{x}_3, \quad \varphi_5 = x_1 x_3, \quad \varphi_6 = x_1 x_2\overline{x}_3, \quad \varphi_7 = x_1 x_2 x_3$$

4.4.4 Hybrid Decision Diagrams

Hybrid Decision Diagrams (HDDs) allow for any of the six matrices in Figure 4.5 to be used for the decomposition of vertices representing a particular variable [33]. Therefore, KBMDs are a subset of HDDs. HDDs were originally investigated for use in reducing the size of DDs for function representation [30]. From a spectral interpretation, these DDs represent the spectrum that would be obtained from a butterfly diagram composed of stages that may be formed using any of the six transformation matrix butterflies specified in Figure 4.5. This can be viewed as a "mixed" spectral transformation.

4.5 EDGE-VALUED DECISION DIAGRAMS

DDs augmented with edge values have gained considerable interest since it has been proven that their use can result in an exponential savings in size for the representation of discrete functions [7]. Edge values are annotations on the edges of DTs or DDs that represent multiplicative or additive factors applied to the subtree being denoted. This section will provide an overview of some common edge-valued DDs and give the underlying spectral interpretation for such graphs.

4.5.1 Edge-valued Binary Decision Diagrams

Edge-Valued Binary DDs (EVBDDs) [105, 106] were introduced to improve the efficiency of MTBDD representations for functions having relatively many different values among the 2^n possible values for the terminal vertices. EVBDDs often provide a smaller representation as compared to MTBDDs. For some classes of functions EVBDDs can result in an exponential savings in computer storage resources [7]. EVBDDs represent discrete functions in the form of a corresponding algebraic polynomial description [149].

In EVBDD representations a nonterminal vertex represents the Boolean decomposition function given in Equation 4.8 and the corresponding pseudo-Boolean function in Equation 4.9. In Equations 4.8 and 4.9, $f_0 = f(x_i = 0)$, $f_1 = f(x_i = 1)$ and v_i represents an additive weighting coefficient associated with the edge corresponding to f_1. It is possible have non-zero additive edge weights for the f_0 cofactor also; however, these are not used in EVBDDs allowing them to be a canonical representation of a function with a given variable order. EVBDDs also have a single terminal vertex representing the constant value of zero.

$$f = x_i(v_i + f_1) + \overline{x}_i f_0 \tag{4.8}$$

$$f = x_i(v_i + f_1) + (1 - x_i)f_0 \tag{4.9}$$

This function can be written as shown in the following.

$$f = f_0 + x_i v_i + x_i(f_1 - f_0) \tag{4.10}$$

Figure 4.9 represents an *Edge-Valued Binary Decision Tree* (EVBDT) for an example function f of $n = 3$ variables. By the application of Equation 4.10 to the vertices of the EVBDT in Figure 4.9, it follows that the tree representing f is in the form as shown in the diagram.

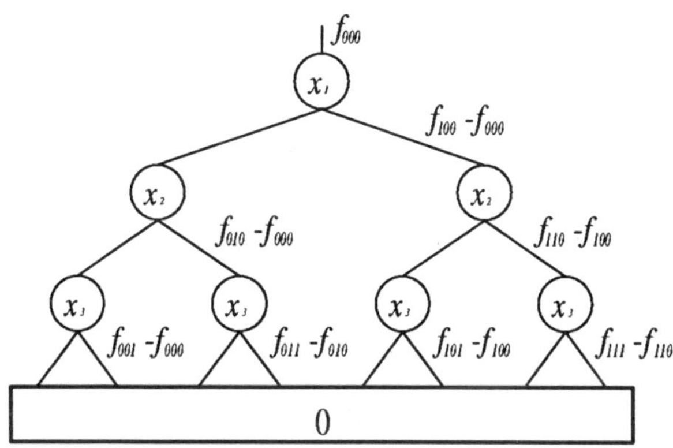

Figure 4.9 General EVBDT for a Function f of $n = 3$ Variables

$$f = f_{000} + x_1(f_{100} - f_{000})$$
$$+x_2(f_{010} - f_{000}) + x_3(f_{001} - f_{000})$$
$$+x_2x_3(f_{011} - f_{010} - (f_{001} - f_{000}))$$
$$+x_1(x_2(f_{110} - f_{100}) + x_3(f_{101} - f_{100})$$
$$+x_2x_3(f_{111} - f_{110} - (f_{101} - f_{100}))$$
$$-x_2(f_{010} - f_{000}) - x_3(f_{001} - f_{000})$$
$$-x_2x_3(f_{011} - f_{010} - (f_{001} - f_{000}))) \tag{4.11}$$

The relationship in Equation 4.11 can be rewritten as given in Equation 4.12.

$$f = f_{000} + x_1(f_{100} - f_{000})$$
$$+x_2(f_{010} - f_{000}) + x_3(f_{001} - f_{000})$$
$$+x_2x_3(f_{011} - f_{010} - f_{001} + f_{000})$$
$$+x_1x_2(f_{110} - f_{100} - f_{010} + f_{000})$$
$$+x_1x_3(f_{101} - f_{100} - f_{001} + f_{000})$$
$$+x_1x_2x_3(f_{111} - f_{110} - f_{101} + f_{100}$$
$$-f_{011} + f_{010} + f_{001} - f_{000}) \tag{4.12}$$

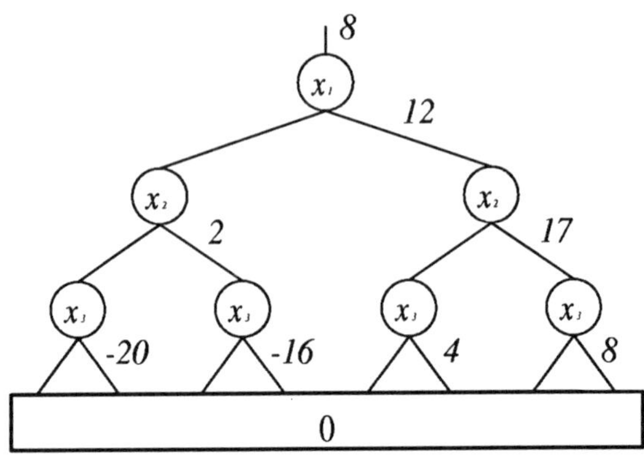

Figure 4.10   EVBDD for Example Function f

Expanding the example function given in Equation 4.5 using the EVBDD decomposition relationship in Equation 4.9 yields Equation 4.13. The zeros in Equation 4.13 correspond to the zero-valued additive weights attributed to the edges that point to f_0 cofactors. The corresponding EVBDD is shown in Figure 4.10.

$$
\begin{aligned}
f \;=\; & 8 + x_1(12 + x_2(17 + x_3(8)) + (1 - x_2)(0 + x_3(4 + 0))) \\
+\; & (1 - x_1)(0 + x_2(2 + x_3(-16 + 0)) + (1 - x_2)(0 + x_3(-20 + 0)))
\end{aligned}
\tag{4.13}
$$

By definition, Equation 4.8 is the algebraic polynomial of f. It is observed that an EVBDD representing some function f is in the form of an algebraic polynomial expression. It follows from Equation 4.11 that the weighting coefficients annotated at edges in EVBDDs are the particular spectral coefficients of partial arithmetic transforms with respect to variables corresponding to the various levels of an EVBDD. Since x is a switching variable, Equation 4.10 can be written as $x(v + l) + \overline{x}r$, where $\overline{x}$ denotes the logical complement of x. It follows by a repeated application of this decomposition type over the nodes of the tree in Figure 4.9, that this tree represents the function in Equation 4.14.

$$\begin{aligned}
f \;=\; & f_{000} + x_1((f_{100} - f_{000}) && (4.14)\\
& + x_2((f_{110} - f_{100}) \\
& + x_3(f_{111} - f_{110} + 0) + \overline{x}_3 \cdot 0) \\
& + \overline{x}_2(x_3(f_{101} - f_{100} + 0) + \overline{x}_3 \cdot 0)) \\
& + \overline{x}_1(x_2((f_{010} - f_{000}) + x_3(f_{011} - f_{010} + 0) + \overline{x}_3 \cdot 0)) \\
& + \overline{x}_2(x_3(f_{001} - f_{000} + 0) + \overline{x}_3 \cdot 0))) \\[6pt]
\;=\; & f_{000} + x_1(f_{100} - f_{000}) + \overline{x}_1 x_2(f_{010} - f_{000}) + x_1 x_2(f_{110} - f_{100}) \\
& \overline{x}_1 \overline{x}_2 x_3(f_{001} - f_{000}) + x_1 \overline{x}_2 x_3(f_{101} - f_{100}) \\
& + \overline{x}_1 x_2 x_3(f_{011} - f_{010}) + x_1 x_2 x_3(f_{111} - f_{110})
\end{aligned}$$

This expression can be understood as the expansion of f relative to the following basis functions

$$\begin{aligned}
\varphi_0 &= 1, & \varphi_1 &= x_1, & \varphi_2 &= \overline{x}_1 x_2, & \varphi_3 &= x_1 x_2, \\
\varphi_4 &= \overline{x}_1 \overline{x}_2 x_3, & \varphi_5 &= x_1 \overline{x}_2 x_3, & \varphi_6 &= \overline{x}_1 x_2 x_3, & \varphi_7 &= x_1 x_2 x_3
\end{aligned}$$

which can be represented by the columns of the matrix $\mathbf{Q}$ given by

$$\mathbf{Q} = \begin{bmatrix}
1 & 0 & 0 & 0 & 0 & 0 & 0 & 0 \\
1 & 0 & 0 & 0 & 1 & 0 & 0 & 0 \\
1 & 0 & 1 & 0 & 0 & 0 & 0 & 0 \\
1 & 0 & 1 & 0 & 0 & 0 & 1 & 0 \\
1 & 1 & 0 & 0 & 0 & 0 & 0 & 0 \\
1 & 1 & 0 & 0 & 0 & 1 & 0 & 0 \\
1 & 1 & 0 & 1 & 0 & 0 & 0 & 0 \\
1 & 1 & 0 & 1 & 0 & 0 & 0 & 1
\end{bmatrix}$$

The matrix $\mathbf{Q}^{-1}$ is the inverse of $\mathbf{Q}$ which is given by

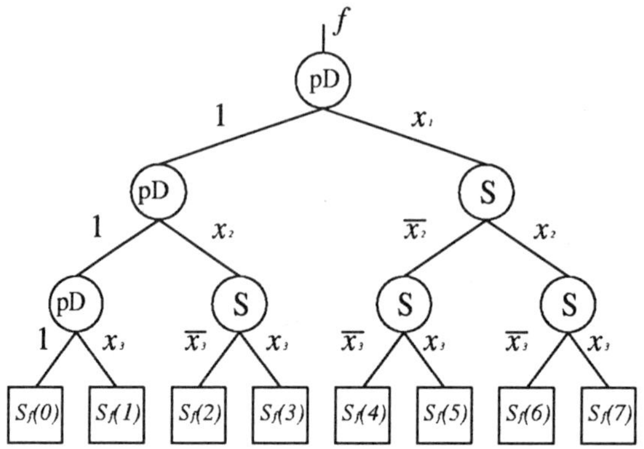

Figure 4.11   A PKDD corresponding to an EVBDD for $n = 3$

$$\mathbf{Q}^{-1} = \begin{bmatrix} 1 & 0 & 0 & 0 & 0 & 0 & 0 & 0 \\ -1 & 0 & 0 & 0 & 1 & 0 & 0 & 0 \\ -1 & 0 & 1 & 0 & 0 & 0 & 0 & 0 \\ 0 & 0 & 0 & 0 & -1 & 0 & 1 & 0 \\ -1 & 1 & 0 & 0 & 0 & 0 & 0 & 0 \\ 0 & 0 & 0 & 0 & -1 & 1 & 0 & 0 \\ 0 & 0 & -1 & 1 & 0 & 0 & 0 & 0 \\ 0 & 0 & 0 & 0 & 0 & 0 & -1 & 1 \end{bmatrix}$$

This is the transform matrix that can be used to determine the coefficients in
Equation 4.14. The DT shown in Figure 4.11 is the graphical representation
of this expansion. The constant nodes contain the spectral coefficients S_f of f
with respect to the basis $\mathbf{Q}$. They are defined by

$$\mathbf{S}_f = \mathbf{Q}^{-1}\mathbf{F}$$

It is clear that this tree is the integer counterpart of some pseudo-Kronecker
tree which can be reduced to a *Pseudo-Kronecker Decision Diagram* (PKDD)
[138].

EVBDDs represent an alternative representation for the integer-valued counterpart of PKDDs with respect to a particular basis that is related to some subset of the transform as defined by the kernel matrices shown in Figure 4.5.

4.5.2   Factored Edge-valued BDDs

Factored EVBDDs (FEVBDDs) [165] are a modification of EVBDDs designed to improve the efficiency of EVBDDs for representations of functions. For FEVBDDs the nonterminal nodes represent the function

$$f = x_i(v_i + factor_i f_1) + (1 - x_i)f_0$$

where $factor_i$ is the multiplicative weight present at the edge corresponding to the subtree represented by f_1. Thus, the chief difference among EVBDDs and FEVBDDs consists of the use of the multiplicative value in conjunction with the additive weights used in FEVBDDs.

The spectral interpretation of EVBDDs discussed previously applies to FEVBDDs as well since the multiplicative weights are defined in terms of additive weights. Thus they can be expressed through a particular series of partial arithmetic transform coefficients as is illustrated in Figure 4.12 for functions with $n = 3$ variables.

For discrete functions with a large number of different integer values in the complete evaluation of the function, the number of nonterminal nodes at the nth level can be very large. Therefore, in FEVBDDs, the multiplicative weights at the nth level are set to zero and additive weights are normalized to equal unity. Due to this fact, the result of these weighting coefficients in the decomposition of f is manifested in the upper levels of the DT. The multiplicative weights in FEVBDDs are defined through these normalizations in terms of the additive weights used in EVBDDs. This technique provides the improved efficiency of FEVBDDs versus EVBDDs, since the nonterminal vertex levels consist of a smaller number of nonterminal nodes. Thus, for FEVBDDs the number of nonterminal nodes compared to the size of an EVBDD is reduced with a tradeoff consisting of an increased number of weighting coefficients that require storage.

From Equation 4.14, the tree in Figure 4.12 represents f as shown in the following equation.

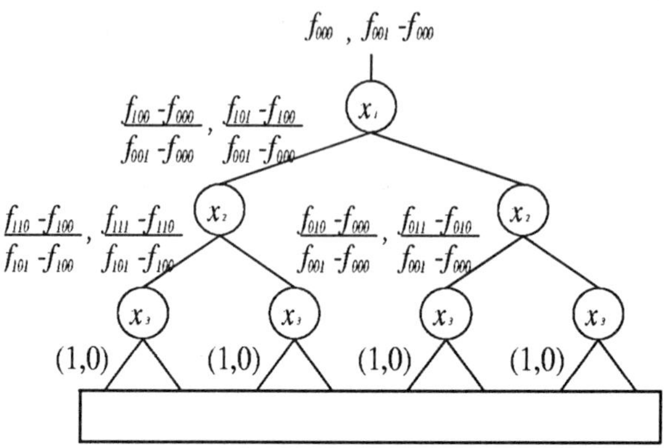

Figure 4.12   Example for a FEVBDD for $n = 3$ Variables

$$f \;=\; A + x_1 E\left[\frac{A}{E} + \left(B + x_2 G\left[\left(\frac{D}{G} + \frac{H}{G}x_3\right) + (1 - x_2)(x_3)\right]\right)\right]$$
$$+\; (1 - x_1)\left[x_2 E\left(\left[\frac{C}{E} + \frac{F}{E}x_3\right] + (1 - x_2)(x_3)\right)\right]$$

Where the following definitions are used.

$$
\begin{aligned}
A &= f_{000}\\
B &= f_{100} - f_{000}\\
C &= f_{010} - f_{000}\\
D &= f_{110} - f_{100}\\
E &= f_{001} - f_{000}\\
F &= f_{011} - f_{010}\\
G &= f_{101} - f_{100}\\
H &= f_{111} - f_{110}
\end{aligned}
$$

After simplification Equation 4.14 reduces to Equation 4.11. Therefore, as for EVBDDs, FEVBDDs represent a function f in the form of an arithmetic polynomial.

DDs derivable in terms of the arithmetic transform (BMDs, EVBDDs and FEVBDDs) are related through the decomposition of f with respect to the integer-valued Reed-Muller functions (or, the arithmetic transforms). For the case of BMDs the coefficients in the decomposition are expressed at the terminal nodes, while for EVBDD representations they are transferred to the edges. In FEVBDD representations the coefficients are moved to the edges at the upper levels in the DD.

4.5.3   *Binary Moment Diagrams

BMDs [18] are the edge-valued version of BMDs. The difference with respect to EVBDDs is in the order of the decompositions applied in the definition of these types of DDs. Additionally, EVBDDs make use of additive edge values while *BMDs use multiplicative ones.

In EVBDD representations the linear decomposition is used to assign f to the binary decision tree. Then the arithmetic transform is used to determine the values assigned to the edges. The partial arithmetic transforms are applied non-recursively to f, since the recursive structure of the DTs is already expressed through the partial linear decompositions in the same way as in MTBDDs.

In *BMD representations, the arithmetic transform is performed to determine the basic DT, and then the linear decomposition is performed to determine the values at the edges. Therefore *BMDs are the edge-valued version of BMDs, where the weighting coefficients at the edges are determined through the linear decomposition of the arithmetic transform coefficients. In this way the common factors in these coefficients are moved to the edges, and the remaining parts are kept as the values of terminal nodes in *BMDs. Therefore the decomposition used in *BMDs is a modification of the decomposition used in BMDs and is given by

$$f = w_i[f_0 + x_i(f_1 - f_0)]$$

where the w_i values are the weighting coefficients. The number of factors allowed in the factorization of the arithmetic transform coefficients is at most equal to the number of levels in the DT, *i.e.*, to the number of variables in f.

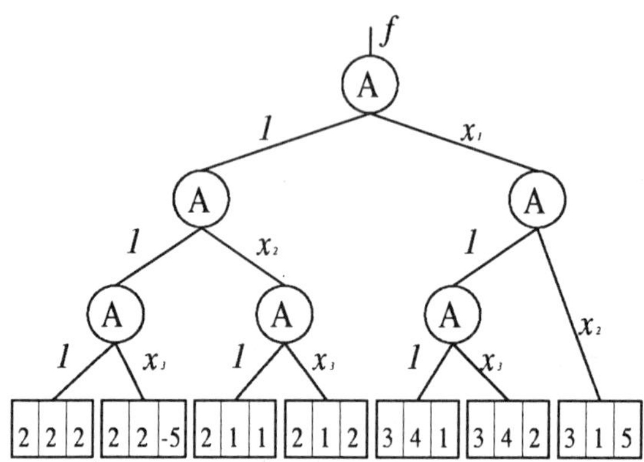

Figure 4.13   BMD with Factored Terminal Values

To illustrate these properties consider the following example taken from [18] where it was used to illustrate the definition of *BMDs and their differences with BMDs.

Example 4.1 *Figure 4.13 contains terminal vertices that are a factored form of the terminal values (or, correspondingly, the arithmetic transform coefficients) present in the BMD shown in Figure 4.7. Many of the factored quantities have the same value. By annotating the 1-edges exiting each nonterminal vertex with their corresponding factor and then performing a DD reduction, the *BMD may be produced as shown in Figure 4.14.*

As is noted in Chapter 3, the arithmetic transform can be calculated through the recursive application of partial arithmetic transforms. Example 4.2 illustrates this principle.

Example 4.2 *Let $\mathbf{A}_i, i = 1, 2, 3$ denote the matrices describing the partial arithmetic transforms for a function of $n = 3$ variables.*

*For the function f in Example 4.1, the calculation of the arithmetic spectrum and the determination of the weighting coefficients in the corresponding *BMD can be performed as follows*

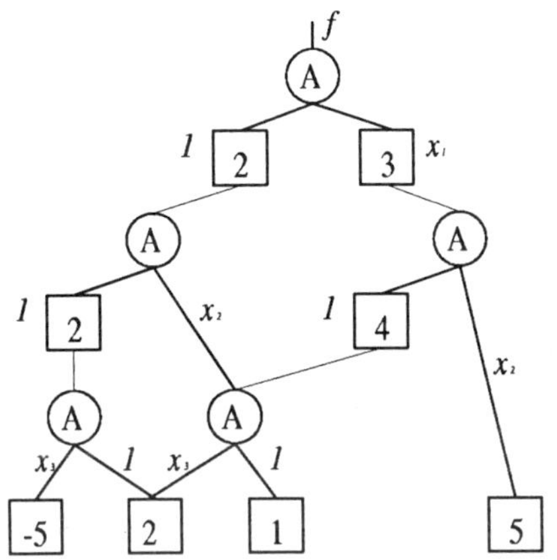

Figure 4.14   *BMD for f in Example 4.1

$$
\mathbf{A}_1(f) =
\begin{bmatrix}
1 & 0 & 0 & 0 & 0 & 0 & 0 & 0 \\
0 & 1 & 0 & 0 & 0 & 0 & 0 & 0 \\
0 & 0 & 1 & 0 & 0 & 0 & 0 & 0 \\
0 & 0 & 0 & 1 & 0 & 0 & 0 & 0 \\
-1 & 0 & 0 & 0 & 1 & 0 & 0 & 0 \\
0 & -1 & 0 & 0 & 0 & 1 & 0 & 0 \\
0 & 0 & -1 & 0 & 0 & 0 & 1 & 0 \\
0 & 0 & 0 & -1 & 0 & 0 & 0 & 1
\end{bmatrix}
\begin{bmatrix}
8 \\ -12 \\ 10 \\ -6 \\ 20 \\ 24 \\ 37 \\ 45
\end{bmatrix}
=
\begin{bmatrix}
2\begin{bmatrix} 4 \\ -6 \\ 5 \\ -3 \end{bmatrix} \\
3\begin{bmatrix} 4 \\ 12 \\ 9 \\ 17 \end{bmatrix}
\end{bmatrix}
$$

$$
\mathbf{A}_2(\mathbf{A}_1(f)) =
\begin{bmatrix}
1 & 0 & 0 & 0 & 0 & 0 & 0 & 0 \\
0 & 1 & 0 & 0 & 0 & 0 & 0 & 0 \\
-1 & 0 & 1 & 0 & 0 & 0 & 0 & 0 \\
0 & -1 & 0 & 1 & 0 & 0 & 0 & 0 \\
1 & 0 & 0 & 0 & 1 & 0 & 0 & 0 \\
0 & 1 & 0 & 0 & 0 & 1 & 0 & 0 \\
-1 & 0 & 1 & 0 & -1 & 0 & 1 & 0 \\
0 & -1 & 0 & 1 & 0 & -1 & 0 & 1
\end{bmatrix}
\begin{bmatrix}
2\begin{bmatrix} 4 \\ -6 \\ 5 \\ -3 \end{bmatrix} \\
3\begin{bmatrix} 4 \\ 12 \\ 9 \\ 17 \end{bmatrix}
\end{bmatrix}
$$

$$= \begin{bmatrix} 2 \begin{bmatrix} 2 \begin{bmatrix} 2 \\ -3 \end{bmatrix} \\ \begin{bmatrix} 1 \\ 3 \end{bmatrix} \end{bmatrix} \\ 3 \begin{bmatrix} 4 \begin{bmatrix} 1 \\ 3 \end{bmatrix} \\ \begin{bmatrix} 5 \\ 5 \end{bmatrix} \end{bmatrix} \end{bmatrix}$$

$$\mathbf{A}_3(\mathbf{A}_2(\mathbf{A}_1(f))) = \begin{bmatrix} 1 & 0 & 0 & 0 & 0 & 0 & 0 & 0 \\ -1 & 1 & 0 & 0 & 0 & 0 & 0 & 0 \\ 0 & 0 & 1 & 0 & 0 & 0 & 0 & 0 \\ 0 & 0 & -1 & 1 & 0 & 0 & 0 & 0 \\ 0 & 0 & 0 & 0 & 1 & 0 & 0 & 0 \\ 0 & 0 & 0 & 0 & -1 & 1 & 0 & 0 \\ 0 & 0 & 0 & 0 & 0 & 0 & 1 & 0 \\ 0 & 0 & 0 & 0 & 0 & 0 & -1 & 1 \end{bmatrix} \begin{bmatrix} 2 \begin{bmatrix} 2 \begin{bmatrix} 2 \\ -3 \end{bmatrix} \\ \begin{bmatrix} 1 \\ 3 \end{bmatrix} \end{bmatrix} \\ 3 \begin{bmatrix} 4 \begin{bmatrix} 1 \\ 3 \end{bmatrix} \\ \begin{bmatrix} 5 \\ 5 \end{bmatrix} \end{bmatrix} \end{bmatrix}$$

$$= \begin{bmatrix} 2 \begin{bmatrix} 2 \begin{bmatrix} 2 \\ -5 \end{bmatrix} \\ \begin{bmatrix} 1 \\ 2 \end{bmatrix} \end{bmatrix} \\ 3 \begin{bmatrix} 4 \begin{bmatrix} 1 \\ 2 \end{bmatrix} \\ \begin{bmatrix} 5 \\ 0 \end{bmatrix} \end{bmatrix} \end{bmatrix}$$

The position of the multiplicative factors in the resulting spectral vector corresponds to the polarity of the variable x_i in the same way as in Figure 4.14.

The arithmetic transform can be calculated through a recursive application of partial arithmetic transforms. Applied recursively, the partial arithmetic

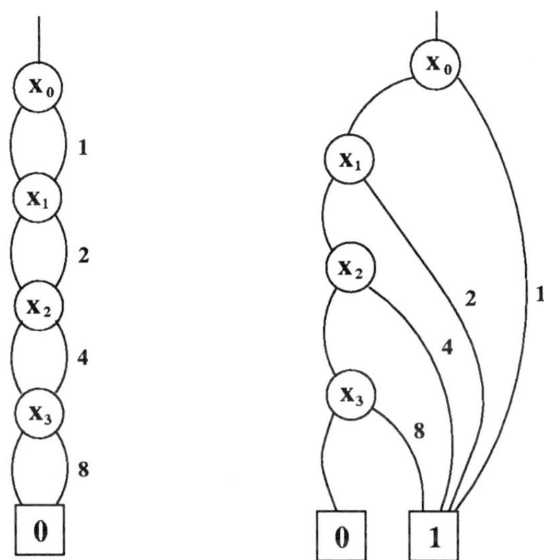

Figure 4.15 EVBDD and *BMD for Unsigned Integer Encoding

transforms may be viewed as steps in "fast" spectral transform algorithms that are described in detail in Chapter 5. Therefore, the linear decomposition used in *BMDs to determine the weighting coefficients, can be performed within this recursive algorithm.

An EVBDD and *BMD for the unsigned integer encoding (4 bits)

$$f_{enc}(x_3, x_2, x_1, x_0) = [x_3, x_2, x_1, x_0] = \sum_{i=0}^{3} 2^i x_i$$

is given in Figure 4.15. The left outgoing edge of each x_i node represents $low(x_i)$. At the DD edges the additive or multiplicative values are displayed. Square vertices represent the leaf nodes. Let g_1, g_2, and g_3 be the functions represented by the nodes for x_1, x_2, and x_3 respectively in the EVBDD. Then the expressions for g_i and f_{enc} are given as follows

$$\begin{aligned}
f_{enc}(x_3, x_2, x_1, x_0) &= x_0(1 + g_1) + (1 - x_0)g_1 \\
&= x_0 + g_1
\end{aligned}$$

$$
\begin{aligned}
g_1 &= x_1(2 + g_2) + (1 - x_1)g_2 \\
&= 2x_1 + g_2
\end{aligned}
$$

$$
\begin{aligned}
g_2 &= x_2(4 + g_3) + (1 - x_2)g_3 \\
&= 4x_2 + g_3
\end{aligned}
$$

$$
\begin{aligned}
g_3 &= x_3(8 + 0) + (1 - x_3)0 \\
&= 8x_3
\end{aligned}
$$

The equations for the *BMD form of DD can be derived in a similar manner.

4.5.4   Kronecker * Binary Moment Diagrams

A DD can be formed by setting the values of the terminal nodes to zero and using additive weighting coefficients at the edges based on various Kronecker transforms in the same way as is done for EVBDDs. When multiplicative edge values are also included in this type of structure, a *Kronecker * Binary Moment Diagram* (K*BMD) is formed. The additive and multiplicative coefficients are determined in the same way as is done for FEVBDDs with the exception that Kronecker decompositions are allowed at the nonterminal vertices. These DDs are referred to as *K*BMDs* [45] and can be considered as edge-valued versions of KBMDs. They may also be envisioned as a generalization of *BMDs.

This viewpoint is justified by the idea of replacing the arithmetic transform used in the formulation of BMDs and *BMDs by the differing transforms used in the formulation of K*BMDs. An important difference between EVBDDs and *BMDs is the way in which values of terminal nodes are used. For EVBDDs the terminal nodes are set to zero, as is also the case for K*BMDs. For *BMDs, however, the values of the terminal nodes are integers. They can be set to the integer value 1, and their values can be interpreted as coefficients that are assigned as graph edge annotations. Edge-valued versions of KBMDs that take advantage of this difference may be defined due to the spectral interpretation of *BMDs discussed previously.

Compared to BMDs, the number of vertices required for $*$BMD representations is reduced by sharing common factors in the arithmetic transform coefficients. The tradeoff for this simplification is the introduction of multiplicative factors that are required as edge annotations. As explained above, these factors may be determined during the recursive calculation of the arithmetic transform. The same method may be repeated for the realization of KBMDs in order to derive an edge-valued version of KBMDs representing the counterpart of a $*$BMD.

These definitions of Kronecker $*$BMDs can also be used for the definition of weighting coefficients determined during the calculation of the partial spectral transforms. Carrying this viewpoint further, the possibility of a dynamic optimization operation for a given f may be possible. A different decomposition rule may be chosen after each step of the calculation of the partial spectral transform in order to get a more compact representation or to replace an integer pD and nD node by an integer Shannon node.

After the first two steps of Example 4.2 (the determination of $A_2(A_1(f))$), it is noted that a constant subvector of order 2 results. Since a terminal vertex with value 1 already exists, the pD vertex in the $*$BMD may be replaced by a vertex obeying the Shannon decomposition, and the function f is thus represented by a Kronecker $*$BMD. Continuing the calculation from Example 4.2 results in the following

$$
\mathbf{I}_3(\mathbf{A}_2(\mathbf{A}_1(f))) = \left[\begin{array}{c} 2\left[\begin{array}{c} 2\left[\begin{array}{c} 2 \\ -3 \end{array} \right] \\ \left[\begin{array}{c} 1 \\ 3 \end{array} \right] \end{array} \right] \\ 3\left[\begin{array}{c} 4\left[\begin{array}{c} 1 \\ 3 \end{array} \right] \\ 5\left[\begin{array}{c} 1 \\ 1 \end{array} \right] \end{array} \right] \end{array} \right]
$$

Figure 4.16 contains a diagram representing the K$*$BMD of the $*$BMD in Figure 4.14. The K$*$BMD uses pD nodes for x_1 and x_2 (these may also be interpreted as DD vertices obeying the arithmetic spectral decomposition) and Shannon nodes for x_3. Compared to the $*$BMD shown in Figure 4.14, this example illustrates a small KBMD does not necessarily produce a smaller sized K$*$BMDs.

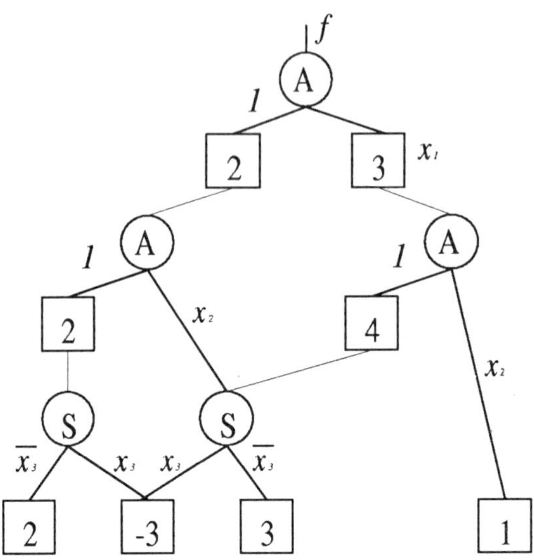

Figure 4.16   K∗BMD for f in Example 4.2

As in the case of ∗BMDs, K∗BMDs may be interpreted from the viewpoint of spectral transforms. Multiplicative factors used as edge annotations are allowed, and the terminal nodes are integer-valued. For the case of K∗BMDs, if the terminal nodes are set to zero and, correspondingly, some of the additive factors at the edges are introduced though partial spectral transforms in the same way as in EVBDDs, the graph representing the counterpart of a corresponding FEVBDD may be derived.

The main difference between the different types of DDs which can be viewed as based upon mixed-spectral transforms (that is, differing 2×2 transformation matrices at each nonterminal vertex) consists in the order of the application of the decompositions which is similar to the difference between EVBDDs and ∗BMDs. For these types of transform-based DDs discussed in [45] and referred to as *K∗BMDs*, the terminal nodes are set to zero and both additive and multiplicative factors are allowed. Based on this fact, these types of DDs can be interpreted as a generalization of ∗BMDs. Depending on the order of decompositions applied to each level of the representative DT which determines the additive and multiplicative coefficients, the underlying representation can also be interpreted as a generalization of FEVBDDs.

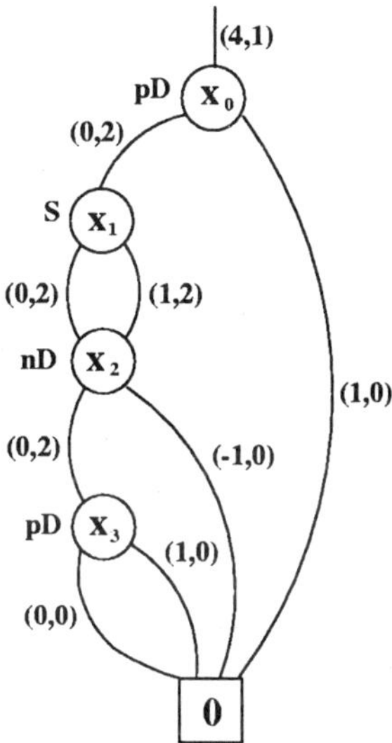

Figure 4.17   K∗BMD Using Unsigned Integer Encoding

Example 4.3 *A K∗BMD using unsigned integer encoding is shown in Figure 4.17. The chosen decomposition types are denoted on the left hand side of each node. At the graph edges the additive and multiplicative values are displayed by (a, m). The decomposition type of each node is also shown.*

K∗BMDs are a generalization of ∗BMDs. This results from the replacement of the arithmetic transform used in BMDs and ∗BMDs by the the pD, nD, or Shannon decompositions used in KBMDs. However, an important difference among EVBDDs and ∗BMDs is the interpretation of the values of the terminal nodes. In EVBDDs the terminal nodes are set to zero (as is also done in the case of K∗BMDs) while in ∗BMDs, the values of terminal nodes are integers. Alternatively, they can be set to one and their values may be interpreted as the coefficients assigned to the edges. An edge-valued version of a KBMD that utilizes this difference may be defined using the spectral interpretation of ∗BMDs.

Compared to BMDs, the number of nodes in ∗BMDs is reduced by sharing common factors in the arithmetic transform coefficients. These factors may be determined during the recursive calculation of the arithmetic transform. The same method may be repeated for KBMDs to derive an edge-valued version of KBMDs representing the proper counterparts of *BMDs.

Such a definition of Kronecker *BMDs with the weighting coefficients determined during the calculation of the partial spectral transforms permits the possibility for dynamic optimization for the representation of a given function f. A different decomposition rule may be chosen after each step of calculation for the partial spectral transform to get a more compact representation or to replace an integer pD or nD node by an integer S node [47, 135].

As in *BMDs, for DDs based on a mixed set of spectral transforms, multiplicative factors at the edges are allowed and terminal nodes are integer-valued. In these DDs if the terminal nodes are set to zero and additive factors at the edges are introduced through the partial spectral transforms in the same way as in EVBDDs, the Kronecker DD representing the counterpart of an FEVBDD is derived.

The main difference between these different extensions of DDs based on spectral transforms consists in the order of the decompositions, similar to the difference between EVBDDs and *BMDs.

It is noted that the K*BMDs discussed earlier have similarities to the definition of FEVBDDs. Because K*BMDs can utilize any of the pD, nD or S decompositions, they may be viewed as a generalization of *BMDs. Depending on the order of decompositions applied to determine additive and multiplicative coefficients, they can also be interpreted as a generalization of FEVBDDs. (For more details on the relationships between the different types of word-level DDs with respect to representation size, see [7].)

4.6 SUMMARY

MTBDDs are a word-level generalization of bit-level BDDs consisting of the integer counterparts to the Shannon nodes. In a spectral interpretation MTBDDs are defined through the decomposition of f in terms of the identity transform I. BMDs are a word-level generalization of bit-level FDDs. They consist of the integer counterparts to pD nodes. In a spectral interpretation, BMDs are defined through decomposition of f in terms of the arithmetic transform.

MTBDDs and BMDs are particular examples of a broad class of word-level Kronecker DDs defined in terms of various spectral transforms. Edge-valued versions of DDs can improve the efficiency of DD representations for functions having exponentially sized Kronecker DDs.

For EVBDDs additive factors are introduced through n decompositions of f in terms of partial arithmetic transforms applied non-recursively to f. For FEVBDDs multiplicative factors are introduced through the normalization of the additive weighting coefficients used in EVBDDs.

Both EVBDDs and *BMDs are defined by the composition of two spectral transforms: the identity transform (as in BDDs and MTBDDs) and the arithmetic transforms (as in BMDs). The differences between these types of DDs consist in the order and the way in which the decompositions are applied to the DT representing f. For *BMDs, the multiplicative factors are introduced through a recursive decomposition of f in terms of the partial arithmetic transforms.

Kronecker versions of these types of DDs may be derived by the replacement of the arithmetic transform by various other spectral transforms with submatrices representing integer counterparts of S, pD and nD nodes.

5

COMPUTATION OF SPECTRAL COEFFICIENTS

The advances in DD representations for discrete valued functions in terms of computational efficiency can be exploited in the calculation of the spectra of Boolean functions. The fundamental matrix approach and the fast transforms derived from the matrix structures were discussed in Chapter 3. This chapter begins with a review of cube list based approaches for computing spectral coefficients after which the computation of spectral coefficients based on output probabilities is considered in some depth. Implementation of all these approaches using DDs is then addressed. A recent method based on Cayley graphs for the computation of the Walsh spectrum is discussed. The chapter concludes with a discussion of incompletely-specified functions and the calculation of their spectra.

5.1 CUBE LIST APPROACHES

The spectrum of a function in its entirety, or a subset of coefficients, may be computed using a *Disjoint Sum of Products* (DSOP) cube list. If the cube list covers the entire function in a disjoint manner, the computation of each coefficient requires each cube to be processed once. This form of spectral computations has been explored in the past in [39, 61, 58, 92, 97, 125, 164].

The underlying theme of this form of spectral transforms is the dependence upon methods to detect DSOP forms for the description of the function. DSOP representations of a given function are not unique; consequently, the derivation of DSOP cube lists has been an area of investigation in the past [13, 58, 102]. In the worst case, a DSOP representation can require 2^{n-1} cubes. As an example, this case occurs for the n-variable parity function $f = x_1 \oplus x_2 \oplus \ldots \oplus x_n$ where each minterm is represented as a single cube.

Various approaches have been developed based on these techniques [39, 58, 125]. In [92] the use of non-disjoint cube lists is examined, and methods for the estimation of the spectra are given as well as techniques for obtaining the exact spectrum.

In the techniques described in [58] the concept of using "standard trivial functions" [90] and comparing $ON(f)$, $OFF(f)$ and $DC(f)$ between the function to be transformed and the trivial functions is used. In the overview of these methods presented here, the idea of using the *output probability* of a function is described [128], and the relationship between this concept and the computation of the spectrum of a Boolean function are given. This approach is general in that the theory of the approach is applicable to computing the spectrum using either DDs or cube lists. A description of output probabilities and their computation is given in Section 2.3.

5.1.1 Computation of the Spectra Using Output Probabilities

This section will describe how output probabilities can be used to compute various spectral coefficients directly. The key idea is to view the computation in terms of a vector-matrix product using an appropriate transformation matrix. Although such products are not computed directly, with this viewpoint each transformation matrix row vector can be considered as an output vector of some other function called a *constituent function*.

By using Boolean relations between the function to be transformed, f, and the various constituent functions, f_c, output probabilities may be computed and used to calculate various spectral coefficients through simple algebraic relations. The advantage of this approach is that the spectral coefficients are computed individually allowing complete storage of 2^n spectral coefficients to be avoided. Furthermore, if efficient methods are used to compute the probability values, a savings in computation time also results.

This relationship is interesting since it allows the global and multi-resolution behavior provided by different forms of spectra and the local behavior specified by the truth table output vector to be related through a common set of values. These relations can be used to develop an efficient means for computing the global Walsh family of spectra, the Reed-Muller spectra and the multi-resolution Haar spectra all based upon the use of output probability computations [152, 157, 160].

5.1.2   Walsh Spectrum

Since the relationships described here apply only to a single Walsh coefficient, the particular coefficient ordering is not important. Each coefficient is described by the particular f_c that corresponds to a transformation matrix row vector regardless of the actual location of the row vector in the matrix. The Walsh coefficient (using S-encoding) is denoted by $S_f(f_c)$ where f denotes the coefficient with respect to function f, and f_c denotes it is the coefficient resulting from the row vector corresponding to constituent (or basis) function f_c.

In the development of this method S-encoding is used. A derivation using R-encoding follows a similar development. The actual calculation of the coefficient is carried out using integer arithmetic. In terms of computing an inner-product of a transformation row vector and an output vector of a function to be transformed, it is easy to see that each scalar product to be accumulated is either +1 or -1 in value. Furthermore, a scalar product of -1 will only occur when the functions f and f_c have different output values for the same set of input variable assignments.

If we consider the equivalence function given by the exclusive-NOR operation (XNOR), a function can be formed whose output is logic-1 if, and only if, both f and f_c are at the same logic value, $\overline{f \oplus f_c}$. Furthermore, if the output probability of this function is computed, we have the percentage of outputs where both f and f_c simultaneously output the same value, $\wp\{\overline{f \oplus f_c}\}$. The corresponding Walsh spectral coefficient can then be obtained by scaling the output probability value by 2^n, where n is the number of variables in f. This is given in Equation 5.1.

$$S_f(f_c) = 2^n[1 - \wp\{f \oplus f_c\}] \tag{5.1}$$

The Walsh coefficient computation in Equation 5.1 depend upon the equivalence relation of the function being transformed and the constituent function. However, the spectral coefficient can also be computed using Boolean operators other than XOR. This is accomplished by exploiting the probability relationships given in Equations 5.2 and 5.3.

$$\wp\{f + f_c\} = \wp\{f \cdot f_c\} + \wp\{f \oplus f_c\} \tag{5.2}$$

$$\wp\{f + f_c\} = \wp\{f\} + \wp\{f_c\} - \wp\{f \cdot f_c\} \tag{5.3}$$

Substituting the relationships in Equations 5.2 and 5.3 into Equation 5.1 yields the following results.

$$S_f(f_c) = 2^n[1 + 2(\wp\{f \cdot f_c\} - \wp\{f + f_c\})] \tag{5.4}$$

$$S_f(f_c) = 2^n[1 + 4\wp\{f \cdot f_c\} - 2\wp\{f\} - 2\wp\{f_c\}] \tag{5.5}$$

$$S_f(f_c) = 2^n[1 - 4\wp\{f + f_c\} + 2\wp\{f\} + 2\wp\{f_c\}] \tag{5.6}$$

The relationships in Equations 5.5 and 5.6 may be simplified since $\wp\{f_c\} = \frac{1}{2}$ is easily shown to be true for all f_c except the 0th ordered coefficient, $S_f(0)$, where $\wp\{0\} = 0$. In addition, the higher ordered Walsh coefficients can be related to the zero ordered coefficient through the expressions given in Equations 5.7, 5.8 and 5.9.

$$S_f(0) = 2^n(1 - 2\wp\{f\}) \tag{5.7}$$

$$S_f(f_c) = 2^n(4\wp\{f \cdot f_c\} - 1) + S_f(0) \tag{5.8}$$

$$S_f(f_c) = 2^n(3 - 4\wp\{f + f_c\}) - S_f(0) \tag{5.9}$$

These relations may be rearranged to compute the output probabilities directly from the Walsh coefficients if desired.

5.1.3 Reed-Muller Spectrum

The Reed-Muller family of spectra consists of 2^n distinct transformations. These are generally classified according to a *polarity number*. The polarity number is used to indicate if a literal is present in complemented or non-complemented form in the generalized Reed-Muller Boolean algebraic expression. In terms of the Reed-Muller spectra, the polarity number can be considered to uniquely define the transformation matrix. The generalized Reed-Muller spectra have been studied and used extensively in the past. [38] and [75] provide background information for this concept.

In relating output probabilities to the Reed-Muller spectra, considerations must be made due to the fact that they are computed over the Galois field $GF(2)$ and the output probabilities are real quantities in the interval [0,1]. An isomorphic relationship is used to address this problem [158]. Like the Walsh spectrum, each of the rows of the Reed-Muller transformation matrix may be viewed as constituent function output vectors. The constituent functions are actually all possible products of literals (with complementation or lack thereof) determined by the polarity number. Therefore, any arbitrary Reed-Muller spectral coefficient (denoted by $M_f(f_c)$) may be computed as given in Equation 5.10.

$$M_f(f_c) = [2^n\wp\{f \cdot f_c\}](mod\ 2) \tag{5.10}$$

Using the probability relationships given in Equations 5.2 and 5.3, alternative relationships can be derived as given in the following.

$$M_f(f_c) = [2^n(\wp\{f\} + \wp\{f_c\} - \wp\{f + f_c\})](mod\ 2) \tag{5.11}$$

$$M_f(f_c) = [2^n(\wp\{f + f_c\} - \wp\{f \oplus f_c\})](mod\ 2) \qquad (5.12)$$

5.1.4 Haar Spectrum

The Haar spectrum [1, 80, 91, 97] has not been utilized as much use as the Walsh family or Reed-Muller transforms described above. This is unfortunate since the Haar spectrum, and in particular the normalized Haar spectrum, can provide very interesting information concerning the correlation of a function with various Shannon decompositions [92]. The Shannon decompositions have received a lot of scrutiny with the recent popularity of DD representations of Boolean functions since the DD form can often be defined in terms of Shannon decompositions. It is likely that further study of the Haar spectrum particularly in the context of decision diagram techniques will lead to interesting and useful results.

The row vectors of the Haar matrix can in fact be described in terms of the AND operation of the function to be transformed and various cubes formed from the literals. This is in contrast to the Walsh and Reed-Muller transformation matrices where the respective f_c functions were defined with total independence of the function to be transformed. The reason for this difference in row vector definition is because the Haar transform does not give totally global information, rather it gives information regarding the correlation of a function with its various cofactors. However, like the previous approaches, each Haar coefficient can be computed as an algebraic relation of various probability values and therefore the Haar spectrum is also directly linked with output probability calculations.

As in the Walsh family of transforms, the output vector of the function to be transformed generally contains integers with -1 representing logic-1 and +1 representing logic-0. With this viewpoint we can define the number of matches between a particular transformation matrix row vector as the number of times the row vector and function vector components are both simultaneously equal to -1 or +1.

By applying the same principles as those used to determine the algebraic relationships between output probabilities and the Walsh family and Reed-Muller transforms, we can formulate the Haar transform relationships. The presence of cofactors in the Haar constituent functions can be handled by using the rela-

Table 5.1 Relationship of the Haar Spectrum and Output Probabilities

SYMBOL	i	n	p_{m0}	p_{m1}
H_0	0	3	$\wp\{f \cdot 0\}$	$\wp\{\overline{f} \cdot \overline{0}\}$
H_1	0	3	$\wp\{f \cdot x_1\}$	$\wp\{\overline{f} \cdot \overline{x}_1\}$
H_2	1	2	$\wp\{f \cdot \overline{x}_1 \cdot x_2\}/\wp\{\overline{x}_1\}$	$\wp\{\overline{f \cdot \overline{x}_1 \cdot \overline{x}_2}\}/\wp\{\overline{x}_1\}$
H_3	1	2	$\wp\{f \cdot x_1 \cdot x_2\}/\wp\{x_1\}$	$\wp\{\overline{f \cdot x_1 \cdot \overline{x}_2}\}/\wp\{x_1\}$
H_4	2	1	$\wp\{f \cdot \overline{x}_1 \cdot \overline{x}_2 \cdot x_3\}/\wp\{\overline{x}_1 \cdot \overline{x}_2\}$	$\wp\{\overline{f \cdot \overline{x}_1 \cdot \overline{x}_2 \cdot \overline{x}_3}\}/\wp\{\overline{x}_1 \cdot \overline{x}_2\}$
H_5	2	1	$\wp\{f \cdot \overline{x}_1 \cdot x_2 \cdot x_3\}/\wp\{\overline{x}_1 \cdot x_2\}$	$\wp\{\overline{f \cdot \overline{x}_1 \cdot x_2 \cdot \overline{x}_3}\}/\wp\{\overline{x}_1 \cdot x_2\}$
H_6	2	1	$\wp\{f \cdot x_1 \cdot \overline{x}_2 \cdot x_3\}/\wp\{x_1 \cdot \overline{x}_2\}$	$\wp\{\overline{f \cdot x_1 \cdot \overline{x}_2 \cdot \overline{x}_3}\}/\wp\{x_1 \cdot \overline{x}_2\}$
H_7	2	1	$\wp\{f \cdot x_1 \cdot x_2 \cdot x_3\}/\wp\{x_1 \cdot x_2\}$	$\wp\{\overline{f \cdot x_1 \cdot x_2 \cdot \overline{x}_3}\}/\wp\{x_1 \cdot x_2\}$

tionship in Equation 2.8 to represent these quantities as output probabilities of the AND of the function to be transformed with its respective dependent literals. Note also that the maximum absolute value of a Haar spectral coefficient varies depending on the order of the coefficient. This is due to the reduction in the size of the range of the constituent functions containing cofactors. Table 5.1 contains symbols for each of the Haar coefficients, H_i; values that indicate the size of the constituent function range, i and n, and probability expressions that evaluate whether the function to be transformed and the constituent function simultaneously evaluate to logic-0, p_{m0}, or evaluate to logic-1, p_{m1}.

Using the notation introduced in Table 5.1, an algebraic equation to compute the value of a Haar spectral coefficient can be written in terms of the output probabilities used to compute p_{m0} and p_{m1}.

$$H = 2^{n-i}[2(p_{m0} + p_{m1}) - 1] \tag{5.13}$$

5.2 METHODS BASED ON BDD PATHS AND DISJOINT CUBES

It is well known that the enumerated paths from the root to the terminal-1 vertex of a BDD represent a disjoint set of covering cubes. Based on this observation, the following approaches utilize similar properties, namely, the disjoint or statistical independence of BDD paths. Alternatively, functions may be represented in the form of cube lists where the cubes represent a DSOP expression. One feature of this approach is that spectral coefficients may be computed in any arbitrary order and subsets of coefficients may be extracted. An overview of the use of DSOP cube lists is presented followed by the use of BDDs for the representation of the Boolean function to be transformed.

Although the entire spectrum of a function may be computed using the DD based algorithms described above, the resulting spectral DDs can become too large for practical manipulation. In these cases it is often times desirable to compute a subset of spectral coefficients. In this section we will review several approaches for computing subsets of spectral coefficients based on the use of DDs and cube lists.

As described in a preceding section, the quantity $\wp\{f\}$ is easily computed using a DSOP representation of $ON(f)$. Also, a previous section has described the fact that the basis functions f_c composing a transformation matrix T may be combined using Boolean operations with the function to be transformed in order to compute spectral coefficients using the concept of output probabilities.

These ideas may be combined to formulate a technique for computing spectral coefficients based on DSOP cube list representations. This technique relies on the representation of the function to be transformed, f, and the function representing a row of the transformation matrix, f_c, to each be represented as a DSOP cube list.

To compute a spectral coefficient the following steps are performed:

1. The DSOP cube lists representing f and f_c are combined using an appropriate Boolean operation

2. The output probability of the resulting DSOP cube list is computed

3. The appropriate algebraic expression (depending upon the transform of interest) is applied, resulting in the spectral coefficient.

As an example, consider the function $f = \overline{x}_1\overline{x}_3 + x_2\overline{x}_3 + x_1\overline{x}_2x_3$. One possible DSOP cube list representation of this function is given in Table 2.1. Consider the computation of the Walsh coefficient $S_f(f_c) = S_f(x_1 \oplus x_3)$. In this case $f_c = x_1 \oplus x_3$. Using Equation 5.7 followed by 5.8 the desired coefficient is computed. The value $\wp\{f\} = 1/2$ is as shown in Table 2.1. Using this value in Equation 5.7, it is seen that the 0th-ordered Walsh coefficient is $S_f(0) = [2^3(1 - 2 \times \frac{1}{2})] = 0$. From Equation 5.8 it is necessary to form the Boolean AND of f_c with the function to be transformed, f. In terms of DSOP cube list operations, this requires ANDing each cube representing f_c with those representing f. The output probability of the resulting cube list is computed, $\wp\{f \cdot f_c\}$, and used. It is noted that the cube list of $f \cdot f_c$ is guaranteed to be in DSOP form since each of the cube list representing f and f_c are DSOP. For this example the resulting cube list consists of the single term, 110, which happens to be a single minterm. Hence, $\wp\{f \cdot f_c\} = \{f \cdot (x_1 \oplus x_3)\} = \frac{1}{8}$. Using Equation 5.8 it is seen that the Walsh spectral coefficient is $S_f(f_c) = 8[4(\frac{1}{8} - 1)] + 0 = -4$.

This type of computation is equally applicable to the computation of the Reed-Muller and Haar spectra. In particular, the Reed-Muller spectra consists of a set of $\{f_c\}$ that are representable as single product terms; hence, the resulting DSOP cube list representing $f \cdot f_c$ is relatively simple to construct. As an example, the Reed-Muller spectral coefficient for a 3-variable function with $f_c = \overline{x}_1 \cdot x_2$ results in the computation of $M_{12} = [8 \times \wp\{f \cdot f_c\}](mod\ 2)$. Using the f in the example given in the previous paragraph, it is seen that $f \cdot f_c = \frac{1}{8}$. Using this result with Equation 5.10 results in the computation $M_{12} = [8 \cdot \frac{1}{8}](mod\ 2) = 1$.

5.3 SPECTRAL COMPUTATION USING DECISION DIAGRAMS

The complete spectrum of a Boolean function consists of 2^n elements that may often be represented in a compact form using the MTBDD data structure as described in the previous chapter. This fact provides the motivation for formulating DD based algorithms that transform the BDD description of a Boolean function into an MTBDD that represents the corresponding spectrum. Such techniques can be formulated using the transform matrix product [34] and the so-called "fast" methods originally developed for the discrete Fourier transform [35] and extended to the MTBDD case [117]. Other methods for the computation of spectra from DD representations are given in [59, 60, 131].

$$\begin{array}{c} j_1 j_0 \\ i_1 i_0 \end{array} \quad \begin{array}{cccc} 00 & 01 & 10 & 11 \end{array}$$

$$\begin{array}{c} 00 \\ 01 \\ 10 \\ 11 \end{array} \left[\begin{array}{cccc} 1 & 1 & 1 & 1 \\ 1 & -1 & 1 & -1 \\ 1 & 1 & -1 & -1 \\ 1 & -1 & -1 & 1 \end{array} \right]$$

Figure 5.1 4×4 Walsh-Hadamard Transform Matrix with Binary Row and Column Indices

5.3.1 Vector-Matrix Computation Using DDs

The representation of matrices as MTBDDs can lead to compact forms of representation for many transforms of interest [34]. Utilizing these representations and associated DD based algorithms for algebraic operations allows for the direct computation of the spectrum of a Boolean function based upon the mathematical definition (the vector-matrix product). The problem of excessive storage requirements can often be avoided when the underlying data structures are DDs rather than vectors and matrices of integers. This is especially true when the function to be transformed and the resulting spectrum are represented with a relatively small number of DD vertices. The MTBDD representing the spectrum can be very small when many same-valued coefficients exist allowing for a large degree of topological isomorphic subgraph sharing to occur.

As an example, the 4×4 Walsh-Hadamard transformation matrix shown in Figure 5.1 can be represented in MTBDD form as shown in Figure 5.2. The non-terminal vertices correspond to the binary representation of the matrix element coordinates denoted by i (the row coordinate) and j (the column coordinate). The number of distinct variables in the matrix MTBDD representation is $2n$ where n is given by the size of the matrix $2^n \times 2^n$. In the case of Figure 5.2, non-terminal vertices are labeled with the binary variables $\{i_1, i_o, j_1, j_0\}$.

A Boolean function f can be represented as an MTBDD with terminal vertices labeled according to either S-encoding or R-encoding. We shall most frequently use S-encoding. The function's spectrum can be computed by multiplying the MTBDDs representing the transform matrix and the Boolean function resulting in a product MTBDD that represents the spectral vector. The resultant spectral vector has non-terminal vertices corresponding to row indices. As an

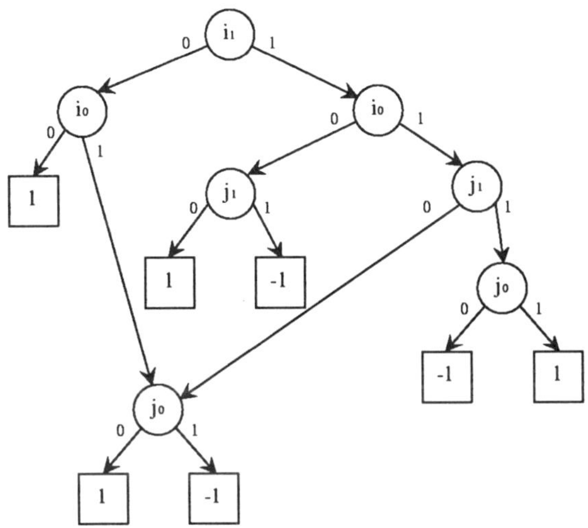

Figure 5.2 MTBDD Representation of 4 × 4 Hadamard-Walsh Matrix

example, the zero-order coefficient is obtained by traversing the all zero path of the MTBDD.

As an example, the MTBDD on the left side of Figure 5.3 depicts the two-input NOR function in the Boolean domain while the MTBDD on the right of the figure represents its Walsh-Hadamard spectrum. It is a peculiarity of this example that the two MTBDDs have the same structure. This is not generally the case and is simply reflexive of the structural simplicity of the NOR function.

For consistency within this book, the MTBDD representing the spectrum of a Boolean function is referred to here as an SDD (as described previously in Chapter 4). The justification for referring to the MTBDD representing a spectrum as an SDD is that the non-terminal vertices of the MTBDD represent a Shannon decomposition while the non-terminal vertices of the SDD can be viewed as corresponding to a different decomposition relationship defined by the particular transformation used. We note that no such distinction was made in [34].

The computation of the product MTBDD (or SDD) representing the spectrum $S(\vec{i})$ is accomplished by using DD based algorithms. This computation is given in Equation 5.14 where $W(\vec{i}, \vec{j})$ represents the transformation matrix, $F(\vec{i})$ represents the function in the Boolean domain, and $\vec{i}$, $\vec{j}$ are vectors of the indices corresponding to row and column elements.

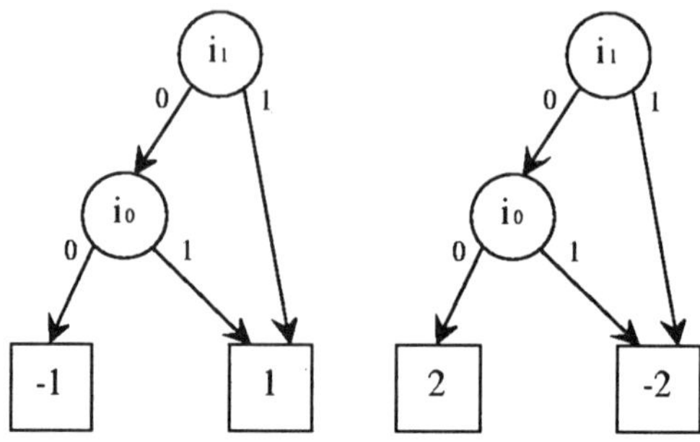

Figure 5.3 MTBDD and Hadamard-Walsh SDD of NOR Function

$$S(\vec{i}) = \sum_{\vec{j}} W(\vec{i}, \vec{j}) F(\vec{i}) \tag{5.14}$$

In terms of implementation of a spectral transformation operation using MTB-DDs, the equation in 5.14 utilizes the property as given in Equation 5.15 and thus is implemented as a series of Boolean AND and MTBDD restriction operations. After each restriction is performed, the elements in $\vec{j}$ are eliminated resulting in the corresponding DD, $S(\vec{i})$, that depends only on the row indices $\vec{i}$.

$$\sum_{\vec{j}} P(\vec{i}, j_1, j_2, \ldots, j_n) \tag{5.15}$$
$$= \sum_{j_1, j_2, \ldots, j_{n-1}} (P(\vec{i}, j_1, j_2, \ldots, j_n, 0) + P(\vec{i}, j_1, j_2, \ldots, j_{n-1}, 1))$$

When a spectral vector contains a large number of different valued coefficients, this technique can result in very large SDDs representing the spectrum since a unique leaf node is required for each different coefficient.

In [71] it was noted that subsets of coefficients can be computed using a reduced transform matrix MTBDD that represents a subset of the rows in the complete transformation matrix. The resulting spectral SDD then represents a subset of spectral coefficients that correspond to the particular transformation matrix rows that were used.

5.3.2   Fast Transform Techniques Using DDs

It is possible to formulate the spectral transformation operation as a graph traversal algorithm resulting in directly converting a BDD into an SDD that represents the spectrum without using graph-based vector-matrix operations [117, 153]. This is accomplished using the one variable, 2×2 transform matrix and traversing the BDD functional representation in either a bottom-up or top-down manner.

In Chapter 3 it was observed that one characterizing factor of a DD is the decomposition type existing at each internal vertex. Given a BDD representing a function to be transformed, it is possible to visit each vertex and to change (transform) the vertex decomposition into a type representing the desired spectrum. When all such vertices have been transformed, the result is an SDD that represents the spectrum.

Since the vertices are transformed individually utilizing the 2×2 transformation matrix, this type of spectrum computation is the graph-algorithm based method of the "fast" transform [35]. Each vertex transformation represents the application of a "butterfly" operation. It is possible that the application of the butterfly operation using a particular DD vertex can lead to adding or removing other vertices from the DD undergoing a transformation. This is to be expected as it is obvious that the BDD in the functional domain is not necessarily topologically similar to the corresponding SDD.

Transformations of Shannon Trees

It is insightful to present the "fast" transformation technique based on DDs by first considering the case of transformations over trees. Although this technique is impractical for implementation, it does allow for the basis of the method to be easily explained.

A Shannon tree is a binary tree representation of a fully-specified single output function. Unlike the BDD representation, Shannon trees are complete and consist of $2^n - 1$ non-terminal vertices and 2^n terminal vertices. The decomposition type at each non-terminal vertex obeys the Shannon decomposition relationship. As an example of the Shannon tree for the three-variable function, $f = \overline{x}_1\overline{x}_3 + x_2\overline{x}_3 + x_1\overline{x}_2x_3$, consider the diagram in Figure 5.4.

Note that S-encoding is used so that the terminal vertices are labeled with integers +1 and -1 as opposed to the corresponding Boolean constant values of 0

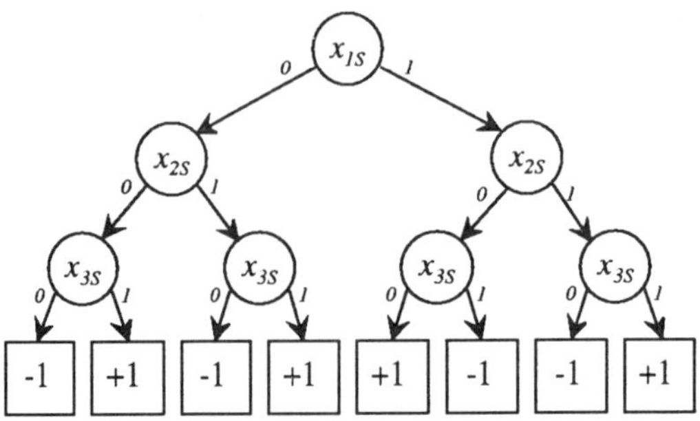

Figure 5.4 Example Shannon Tree with Integer-Valued Terminal Vertices

and 1. Consider the Walsh transform with Hadamard ordering. This transform can be represented in terms of Kronecker products [92]. Note that in terms of graph operations, the one variable transform simply replaces the subtree $low(v)$ with that of a subtree representing their sum and correspondingly replaces the subtree $high(v)$ with that of their difference. This is the butterfly operation for the Walsh transform [146] as phrased in terms of graph operations and is in fact the decomposition relation that replaces the Shannon decomposition. It is easy to see this is the case upon examination of Equation 5.16.

$$\begin{bmatrix} low(v_W) \\ high(v_W) \end{bmatrix} = \begin{bmatrix} +1 & +1 \\ +1 & -1 \end{bmatrix} \begin{bmatrix} low(v_S) \\ high(v_S) \end{bmatrix} \tag{5.16}$$

In terms of implementation, computing the sum (or difference) of two MTBDDs is typically implemented as a recursive procedure similar to the classic `ite` operation used in most BDD package implementations [12].

After the butterfly operation is performed on the Shannon decomposed vertex v_S, it is transformed into the Walsh decomposed vertex v_W. The variable ordering in the tree is important since each butterfly operation may only be applied to the subtrees that have already been transformed. Initially, transforming the terminal vertices to the integers ± 1 allows for the non-terminal nodes at the bottom of the tree to be transformed. By successively applying the transformations in a bottom-up manner, the Shannon tree is transformed into a Walsh tree and the Walsh spectrum is present in Hadamard order from left to right. This method effectively replaces the Shannon decompositions of the non-terminal nodes in the MTBDD with Walsh decompositions. When the decomposition types are replaced, the subtrees rooted at the node undergoing the transformation generally change topology. Figure 5.5 illustrates the state of the tree at various points in the application of the transformation from the Shannon to the Walsh decomposed tree. The vertices drawn with dashed lines indicate portions of the graph that have undergone the transformation. A nonterminal vertex drawn with a dashed circle indicates that the decomposition relation is the Walsh "butterfly", and terminal vertices drawn with a dashed square denote that the corresponding attributed numerical value is a spectral coefficient of the function cofactor represented by the subtree. When the entire graph contains all vertices drawn with dashed lines, the terminal values are spectral coefficients of the original Boolean function.

The analogy between this type of transformation and the commonly known fast transform butterfly diagrams is shown in Figure 5.6 where the Shannon tree is shown with the corresponding butterfly diagram rotated 90° and appearing under the tree. The butterfly diagrams represent signal flow graphs with solid lines indicating a multiplicative factor of +1 and dashed lines represented a factor of -1. The point at which two lines in the subgraph meet indicate the presence of an addition. Figure 5.7 depicts the numerical interpretation of this operation, and the term "butterfly" refers to the fact that the shape of the flowgraph is similar to a butterfly.

This technique can be stated in a succinct form as a depth-first algorithm as given in Figure 5.8 where *Value()* is a function which returns the value of a terminal node, *Index()* is a function which returns the label of a nonterminal node, and *New_Terminal()* and *New_Nonterminal()* are procedures which produce new nodes of the specified types.

DD Algorithms for Fast Transformations

The tree-based algorithm offers no computational advantage over the direct computation of the spectrum using matrix algebra since the size of the tree is exponential in the number of dependent function variables. In order to take advantage of shared topological isomorphic subgraphs as are found in reduced DD structures, the tree-based algorithm must be modified to account for the case when non-terminal variables are present along a path without subsequent valued level indices. This case never occurs in a tree but often does occur in a reduced DD. As an example, consider the case where the function $f = \overline{x}_1 \overline{x}_3 + x_2 \overline{x}_3 + x_1 \overline{x}_2 x_3$ is to be transformed to the Walsh domain. Figure 5.9 contains a diagram representing the reduced BDD of this function with variable order $x_1 \prec x_2 \prec x_3$. As is easily seen, the path specified by $x_1 = 0$ and $x_3 = 0$ skips the intermediate variable x_2. This occurs due to the fact

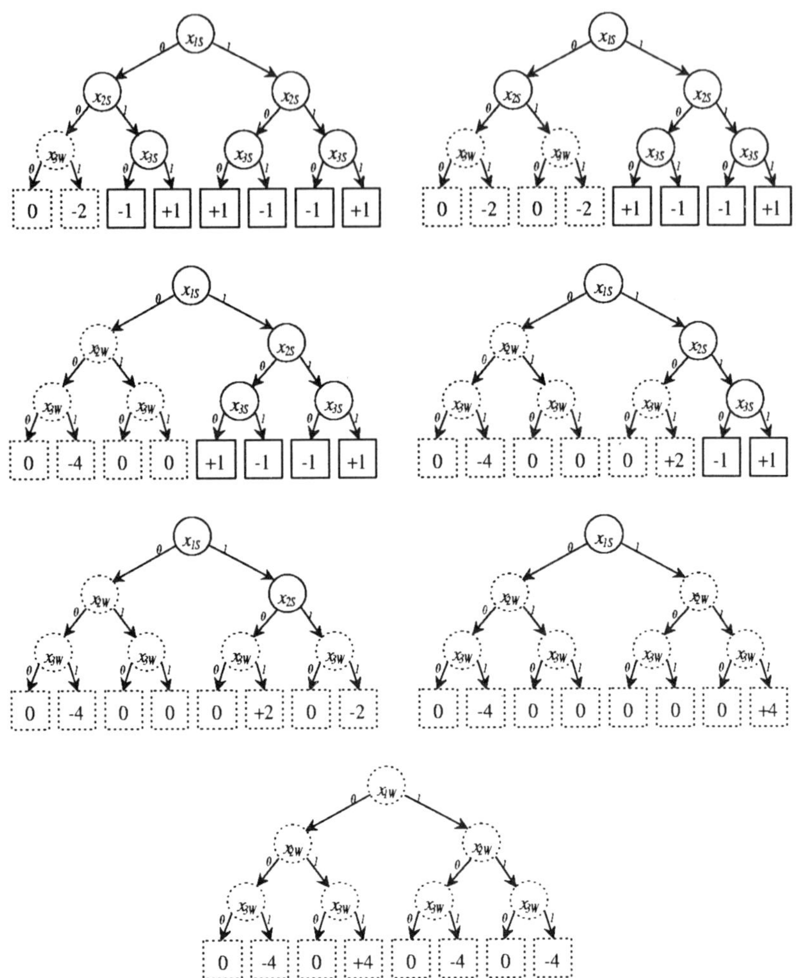

Figure 5.5   Series of Mixed Shannon/Walsh Trees

that x_2 is redundant in this case because both 0-cubes $\overline{x}_1\overline{x}_2\overline{x}_3$ and $\overline{x}_1 x_2\overline{x}_3$ are members of the set $ON(f)$ and are present in the reduced BDD as the path representing the merged cube $\overline{x}_1\,\overline{x}_3$. However, in modifying the decomposition of non-terminal vertex x_1 from a Shannon to a Walsh type, the absence of a vertex representing variable x_2 cannot be ignored and must be inherently considered.

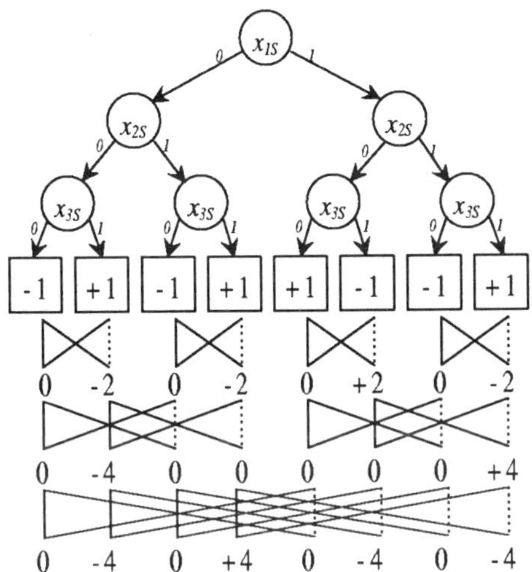

Figure 5.6 Example of Shannon Tree Undergoing a Walsh Transformation with Butterfly Diagram

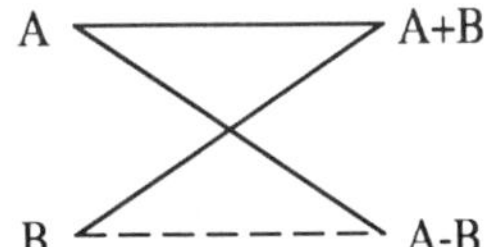

Figure 5.7 Numerical Interpretation of a "Butterfly" Signal Flowgraph for the Walsh Transform

Consider a portion of the BDD shown in Figure 5.9, in particular, the part of the BDD corresponding to the path $x_1 = 0$ and $x_3 = 0$. In Figure 5.10 a subscript is used with each variable to indicate if the corresponding vertex is decomposed using the Shannon or Walsh relationships. Initially, the bottom vertex labeled with variable x_{3S} in Figure 5.9 undergoes a transformation yielding the vertex x_{3W} as shown in the upper, leftmost DD in Figure 5.10. This graph is a hybrid DD (*i.e.* partially MTBDD and partially SDD) consisting of both Shannon and Walsh vertices. In fact, the subtrees rooted in vertices that represent Walsh decompositions correspond to the SDD of Shannon cofactors of the overall function. In this case the bottom vertex labeled x_{3S} in Figure 5.9 points directly to terminal vertices, and the transformation is performed in the same manner as is done for the tree-based approach.

```
Walsh_Tree_Transform (f)
  if(f is a terminal) return
  Walsh_Tree_Transform(Low(f))
  Walsh_Tree_Transform(High(f))
  low_temp = Tree_Addition(Low(f),High(f))
  High(f) = Tree_Subtraction(Low(f),High(f))
  Low(f) = low_temp

Tree_Addition(g,h)
  if(g is a terminal) return(New_Terminal(Value(g)+Value(h))
  return(New_Nonterminal(Index(g),Tree_Addition(Low(g),Low(h)),
                    Tree_Addition(High(g),High(h))))

Tree_Subtraction(g,h)
  if(g is a terminal) return(New_Terminal(Value(g)-Value(h))
  return(New_Nonterminal(Index(g),Tree_Subtraction(Low(g),Low(h)),
                    Tree_Subtraction(High(g),HIgh(h))))
```

Figure 5.8   Pseudo-code for Tree-based Walsh Transformation

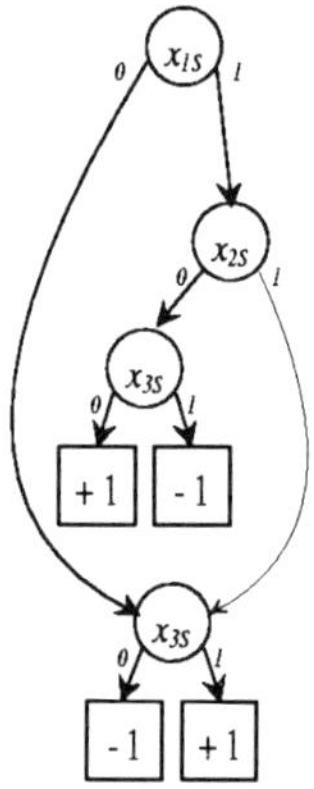

Figure 5.9   Reduced BDD for the Example Function

Next the top vertex x_{1S} in Figure 5.9 must be transformed. Notice that the level value of x_{1S} differs by that for x_{3W} by more than one (*i.e.* $3 - 1 = 2$) indicating that intermediate variables in the ordering are not present in the DD. The "skipped" vertex representing x_{2S} must be considered. By applying the transformation to the "skipped" x_{2S}, we see that the resulting DD does indeed contain vertex x_{2W}. Furthermore, this case can be stated as a general decomposition rule as originally described in [117]. Whenever a path contains two adjacent vertices with level values that differ by two or more, the "skipped" variable v must be inserted into the resulting Walsh SDD with $high(v)$ pointing to terminal value 0 and $low(v)$ pointing to the original subgraph multiplied by two. This occurs since the Walsh decomposition when applied to two identical subtrees, G, results in $G + G = 2G$ and $G - G = 0$. Similar transformation rules apply to other spectral transforms whose transformation matrices can be formulated recursively using Kronecker matrix product operation.

As an example, the function under consideration above can be completely transformed by undergoing a series of vertex transformations during a graph traversal as shown in Figure 5.10. The tree based algorithm outlined above must be modified to operate over MTBDD structures by adding a check for non-sequential index values in a path during the graph traversal and then performing the appropriate transformation.

The algorithm for transforming a BDD is given in Figure 5.11 where for ease of explanation we assume the labels of the nonterminals are ordered increasingly from the root of the BDD towards the terminals and *Index()* applied to a terminal yields a maximum value. *Twice()* doubles the terminal values of the argument BDD an operation which is easily accomplished if edge-valued DD are used.

This algorithm can also be adapted for the Reed-Muller family of transforms. For the positive polarity Reed-Muller transformation [75], Equation 5.17 is used to generate the transform matrix. Equation 5.17 corresponds to the "butterfly" operation for the Reed-Muller transformation, and the corresponding signal flowgraph is shown in Figure 5.12. The matrix multiplication is performed (mod2).

$$\begin{bmatrix} low(v_R) \\ high(v_R) \end{bmatrix} = \left(\begin{bmatrix} +1 & 0 \\ +1 & +1 \end{bmatrix} \begin{bmatrix} low(v_S) \\ high(v_S) \end{bmatrix} \right) (mod\ 2) \qquad (5.17)$$

As an example of this method utilizing the Reed-Muller, positive polarity transform, consider the Boolean function $f = \overline{x}_1\,\overline{x}_3 + x_2\overline{x}_3 + x_1\overline{x}_2x_3$. The Shannon tree representation with the butterfly diagram rotated 90° is shown in Figure

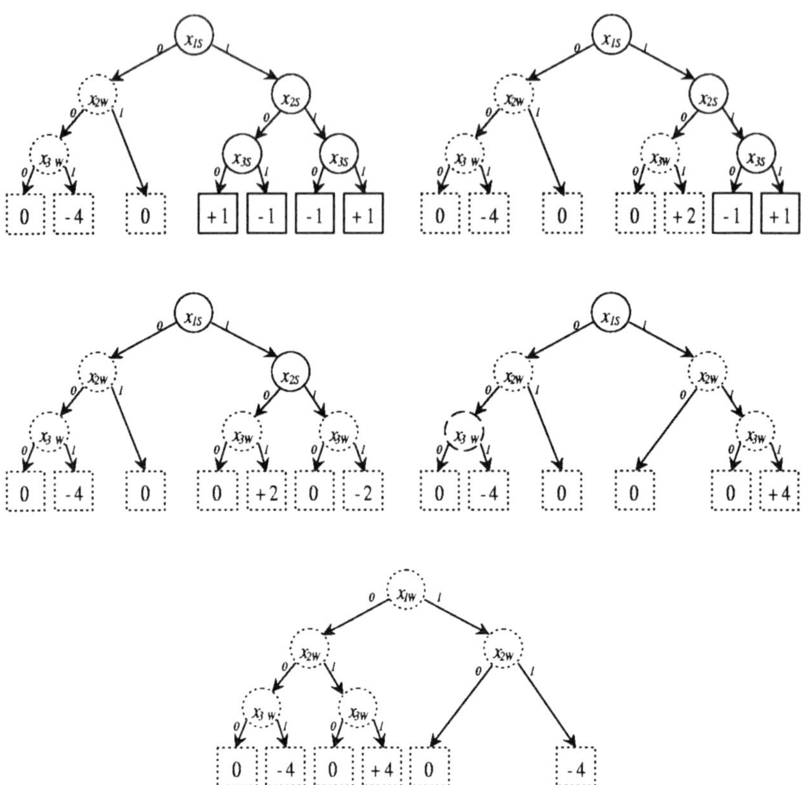

Figure 5.10 BDD Undergoing Walsh Transformation for the Example Function

5.13. As is easily computed by direct transformation into the positive polarity, Reed-Muller domain, this function is equivalent to $f = 1 \oplus x_1 \oplus x_3 \oplus x_1 x_2$.

Figure 5.14 illustrates the application of the tree based algorithm in converting a Shannon tree into the corresponding Reed-Muller spectral tree.

The use of the addition operation over GF(2) during the application of each vertex transformation can be avoided until all vertices have been transformed as is shown in [160]. The alternative is to perform all multiplications using integer arithmetic and then to perform a $(mod\ 2)$ operation mapping each terminal vertex value to 0 or 1 after all decomposition types have been converted to Reed-Muller forms. Although this implementation could possibly simplify intermediate arithmetic computations, the savings is at the expense of potentially

```
Walsh_BDD_Transform (f)
  if(f is a terminal) return
  if(f has already been transformed) return
  Walsh_BDD_Transform(Low(f))
  Walsh_BDD_Transform(High(f))
  low_temp = BDD_Addition(Low(f),High(f))
  High(f) = BDD_Subtraction(Low(f),High(f))
  Low(f) = low_temp


BDD_Addition(g,h)
  if(g and h are terminals)
              return(New_Terminal(Value(g)+Value(h))
  if(Index(g)=Index(h))
              return(New_Nonterminal(Index(g),
                    BDD_Addition(Low(g),Low(h)),
                    BDD_Addition(High(g),High(h))))
  else if(Index(g)<Index(h))
              return(New_Nonterminal(Index(g),
                    BDD_Addition(Low(g),Twice(h)),
                    High(g))
  else return(New_Nonterminal(Index(h),
              BDD_Addition(Low(h),Twice(g)),
              High(h))


BDD_Subtraction(g,h)
  if(g and h are terminals)
              return(New_Terminal(Value(g)-Value(h))
  if(Index(g)=Index(h))
              return(New_Nonterminal(Index(g),
                    BDD_Subtraction(Low(g),Low(H)),
                    BDD_Subtraction(High(g),High(h))))
  else if(Index(g)<Index(h))
              return(New_Nonterminal(Index(g),
                    BDD_Subtraction(Low(g),Twice(h)),
                    High(g))
  else return(New_Nonterminal(Index(h),
              BDD_Subtraction(Low(h),Twice(g)),
              High(h))
```

Figure 5.11 Pseudo-code for BDD-based Walsh Transformation

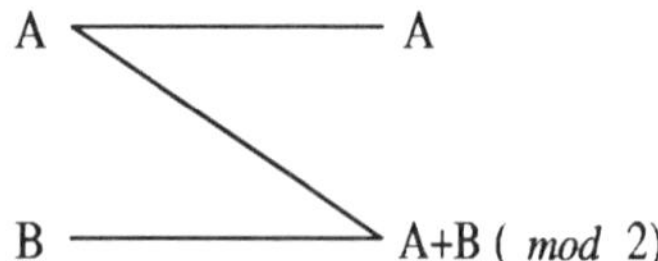

Figure 5.12 Numerical Interpretation of a "Butterfly" Signal Flowgraph for the Positive Polarity Reed-Muller Transform

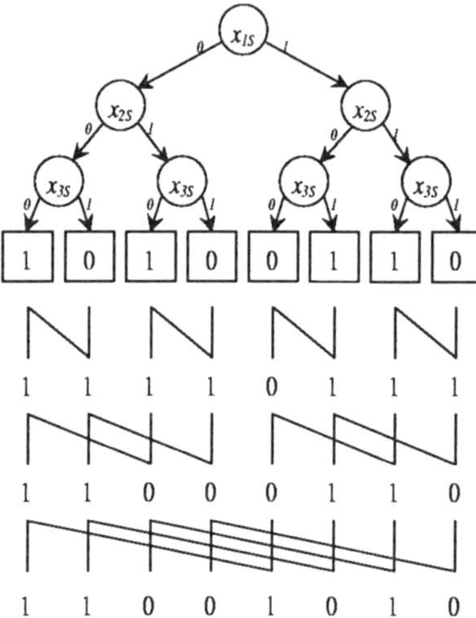

Figure 5.13 Example of Shannon Tree Undergoing a Positve Polarity Reed-Muller Transformation with Butterfly Diagram

creating larger intermediate DDs since more than two terminal vertex values can result. Figure 5.15 illustrates the Shannon tree undergoing a positive polarity Reed-Muller transformation with intermediate integer valued operations and the (*mod* 2) operation applied to the terminal vertices as a final step. In fact Figure 5.15 is the application of the adding transform followed by the (*mod* 2) operation; the arithmetic transform can also be used.

In a similar manner to the development of the Walsh transform discussed above, a depth-first, bottom-up algorithm can be developed and implemented as graph operations over a BDD. In terms of the tree based algorithm, the only change to the algorithm for computation of the Walsh tree is that the butterfly function is changed to obey the relationship given in Equation 5.17. The extensions to the DD based algorithm are straightforward, and Figure 5.16 illustrates the state

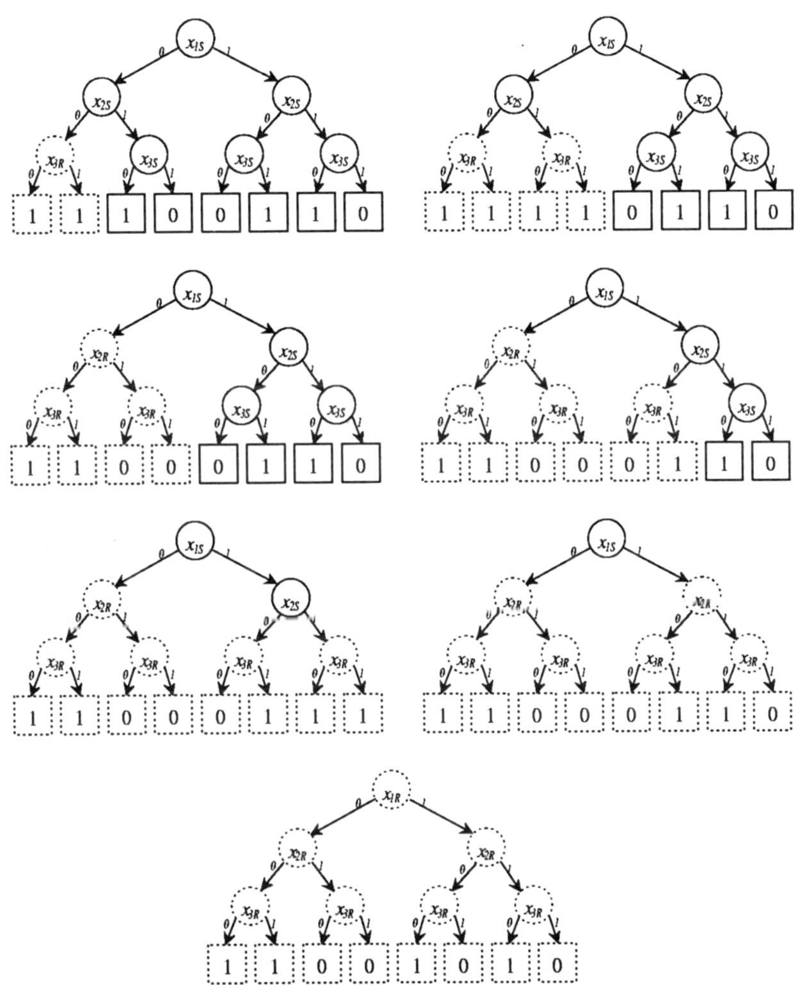

Figure 5.14 Series of Shannon/Reed-Muller Undergoing a Positve Polarity Reed-Muller Transformation

of the BDD as it undergoes transformation to a Reed-Muller, positive polarity SDD.

In general, all 2^n possible fixed polarity Reed-Muller transforms can be implemented using this technique. The only modification required is that variables appearing in complemented form must undergo the negative polarity Reed-

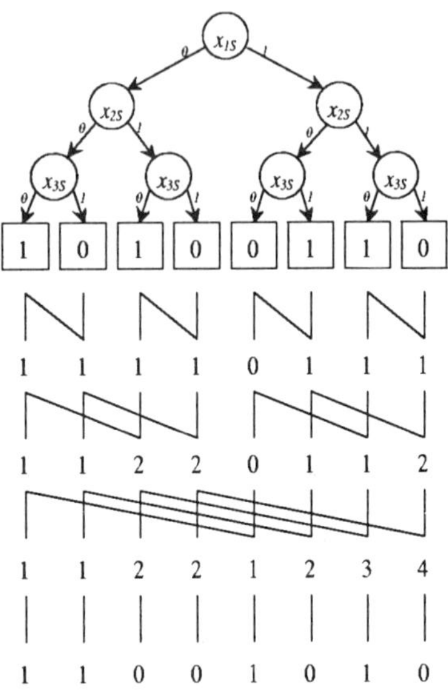

Figure 5.15 Example of Shannon Tree Undergoing a Positve Polarity Reed-Muller Transformation with Integer Operations

Muller decomposition. This relationship is given in Equation 5.18, and the corresponding signal flowgraph is depicted in Figure 5.17.

It is noted that the SDDs for various fixed polarity Reed-Muller transforms are a different interpretation of the *Functional Decision Diagrams* (FDDs) as originally developed in [42]. The decomposition relations in Equations 5.17 and 5.18 are also known as the *positive-Davio* (pD) and *negative-Davio* (nD) decomposition types respectively.

$$\begin{bmatrix} low(v_R) \\ high(v_R) \end{bmatrix} = \left(\begin{bmatrix} +1 & +1 \\ 0 & +1 \end{bmatrix} \begin{bmatrix} low(v_S) \\ high(v_S) \end{bmatrix} \right) (mod\ 2) \tag{5.18}$$

The Haar transform can also be implemented in a similar manner; however, the tree based algorithm differs from the case used for the Walsh and Reed-Muller transforms. The difference arises due to the "shape" of the spectral transform signal flowgraph. In the case of the Walsh and Reed-Muller transforms, there are $n2^{n-1}$ butterfly operations required. For the Haar transform the total number of butterfly operations is reduced and is equal to $\sum_{i=1}^{n} 2^{n-i}$. To illustrate

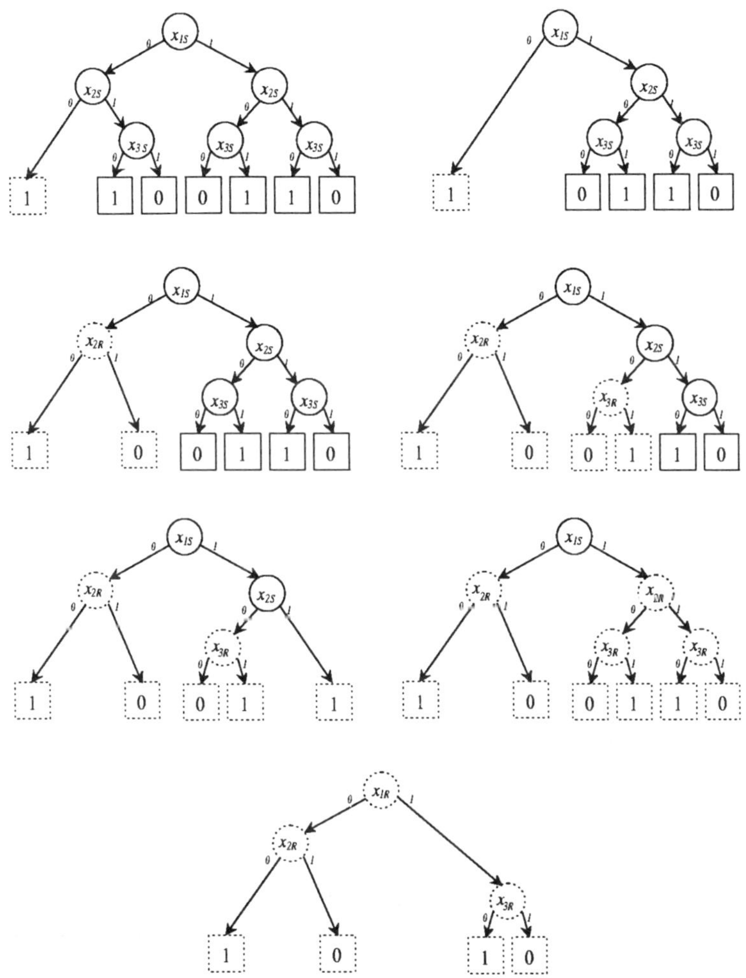

Figure 5.16 BDD Undergoing Positive Polarity Reed-Muller Transformation for the Example Function

the reduced number of butterfly operations, consider Figure 5.18 where the Shannon tree for the Boolean function $f = \overline{x}_1\overline{x}_3 + x_2\overline{x}_3 + x_1\overline{x}_2x_3$ is shown with the corresponding butterfly diagram beneath.

The tree based algorithm for this transformation may be stated in pseudo-code form as given in Figure 5.19. As an example, Figure 5.20 illustrates the transformation of the Shannon tree shown in Figure 5.18 into a corresponding

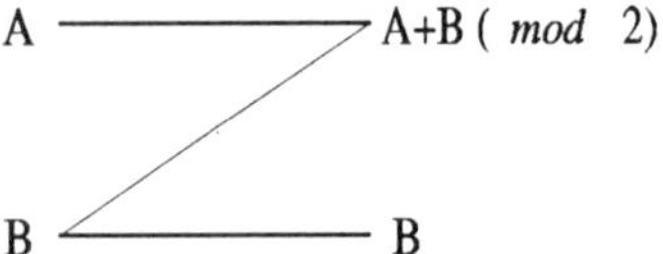

Figure 5.17 Numerical Interpretation of a "Butterfly" Signal Flowgraph for the Negative Polarity Reed-Muller Transform

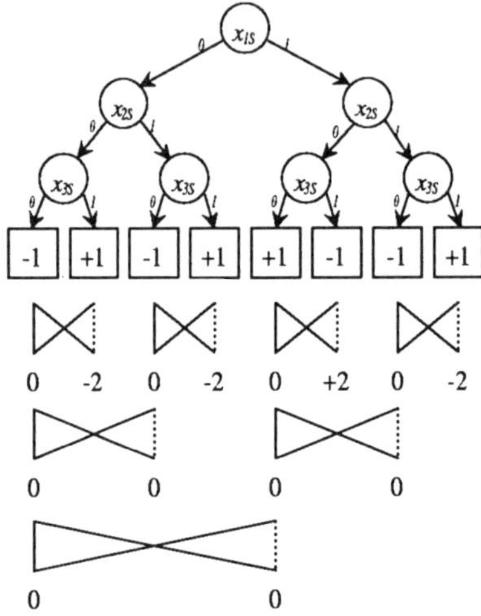

Figure 5.18 Example of Shannon Tree Undergoing a Haar Transformation

Haar spectral tree. The algorithm in Figure 5.19 can be extended to the BDD case in a manner similar to that shown in Figure 5.11 for the Walsh case.

By incorporating checks to determine if BDD variable levels are skipped, the tree-based algorithm can also be written as a BDD based transformation algorithm. Note that the Haar butterfly is the same as that for the Walsh transformation. The only difference is that some butterfly operations are missing. This also has the desirable effect of generating some coefficients before the computation is complete: the so-called high-resolution coefficients. As an example of the BDD based algorithm, a series of hybrid BDD/SDDs are shown in Figure 5.21.

```
Haar_Tree_Transform (f)
   if(f is a terminal) return
   Haar_Tree_Transform(Low(f))
   Haar_Tree_Transform(High(f))
   low_branch = Low(f)
   while(low_branch not a terminal)
     low_branch = Low(low_branch)
   high_branch = High(f)
   while(high_branch not a terminal)
     high_branch = Low(high_branch)
   temp_value = Value(low_branch)+Value(high_branch)
   Value(high_branch) = Value(low_branch)-Value(high_branch)
   Value(low_branch) = temp_value
```

Figure 5.19   Pseudo-code for the Tree-based Haar Transformation

Edge Attributed DDs

It is noted that the use of attributed edges may also be incorporated into these algorithms resulting in more compact DDs and, hence, reduced transformation computation time. As an example, negative-edge attributes in the BDD representation of the function can be extended into arithmetic sign attributes for the SDD. Also, edge-values can be used to further reduce the size of the SDD allowing it to be represented as a BMD or hybrid MTBDD/BMD as described in [33]. In [83] it was shown that the SDD with edge attributes containing **Haar** spectral coefficients is isomorphic to the **BDD** representing the function in the Boolean domain.

5.4   COMPUTATION BASED ON CAYLEY GRAPHS

An alternative approach for the computation of the Walsh spectrum of a Boolean function based on algebraic groups and graph theory is described in [8]. The Cayley graph (or Cayley color graph) is a structure that is used to relate an algebraic group to graph theory [22, 23, 169]. This technique for the computation of the Walsh spectrum of some function f relies upon representing the Boolean function based upon a specific definition of a group as given in [8].

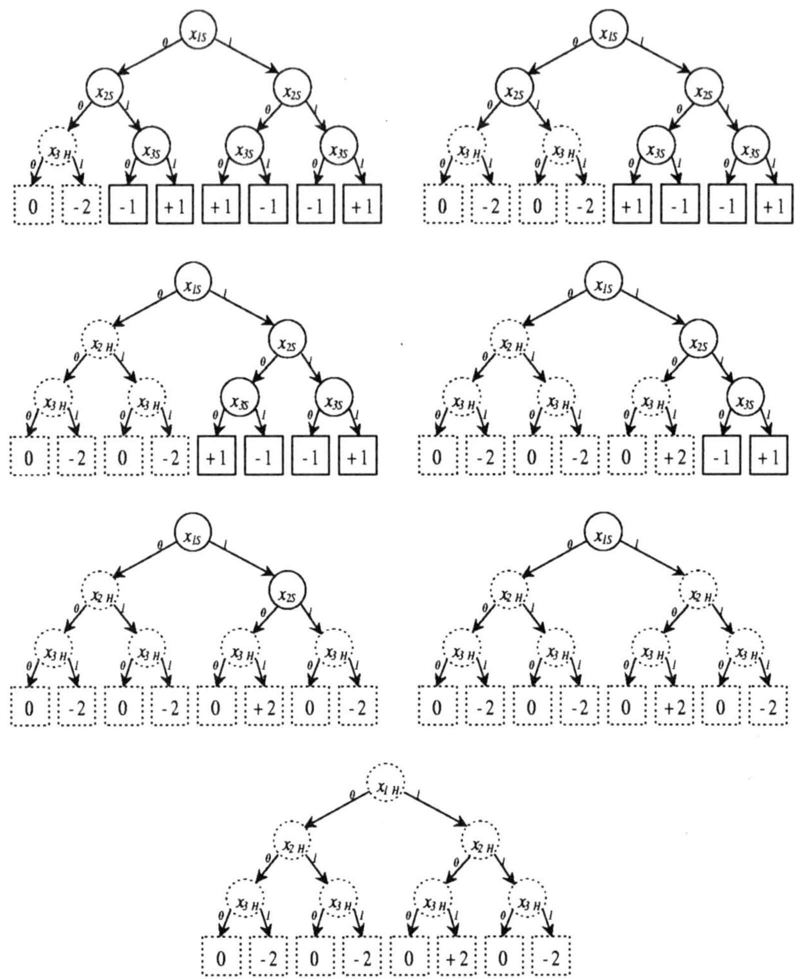

Figure 5.20 Shannon Tree Undergoing a Haar Transformation for the Example Function

Recall that a group $(G, *)$ consists of a set G and a binary operator $*$ such that the operation $*$ over all $x \in G$ is associative, that there is an element $e \in G$ that serves as an identity for all $x \in G$ based on $*$, and that for all $x \in G$ there exists some $x^{-1} \in G$ such that $x * x^{-1} = e$ (inverses exist).

The particular group used in this technique $(M, \oplus)$ characterizes the Boolean function $f : B^n \to B$. $(M, \oplus)$ is a group where M consists of all possible

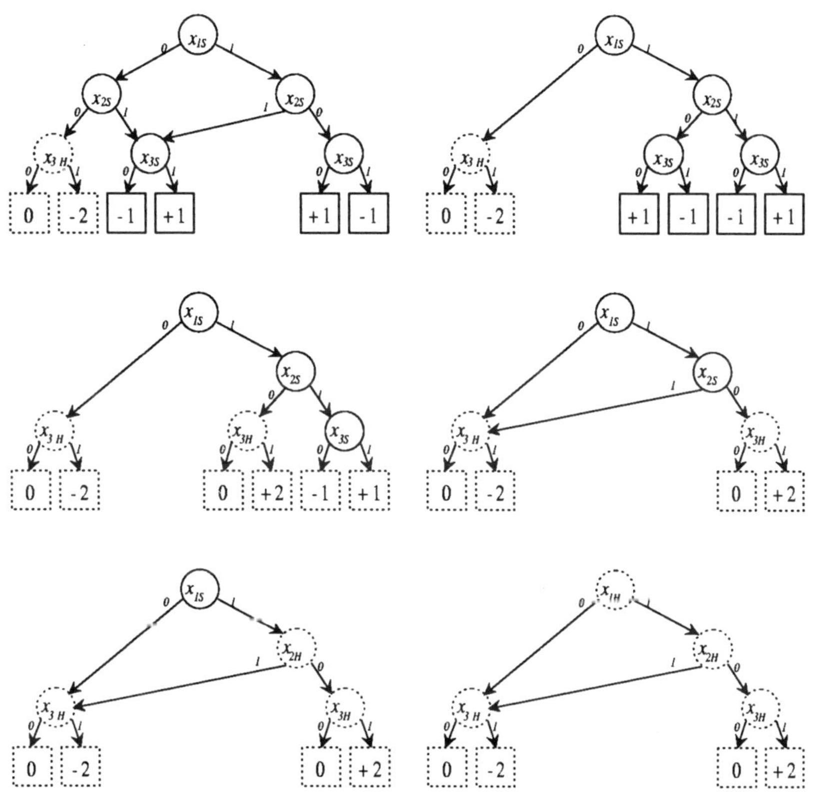

Figure 5.21 BDD Undergoing Haar Transformation for the Example Function

minterms in B^n (that is, all points in the space defined by B^n) and $\oplus$ is the binary operator. Using the cube notation defined earlier, it is easily seen that this group has an identity element corresponding to an n-length bit string of all zeros. Additionally, for each element $m_i \in M$, $m_i^{-1} = m_i$.

The Cayley graph $G_{Cay}(V, E)$ corresponding to this group representing the Boolean function f has a vertex set V where each $v_i \in V$ uniquely corresponds to an element of the set $m_i \in M$. The edge set E is given by Equation 5.19 below.

$$E = \{(m_i, m_j) \in B^n \times B^n | f(m_i \oplus m_j) = 1\} \tag{5.19}$$

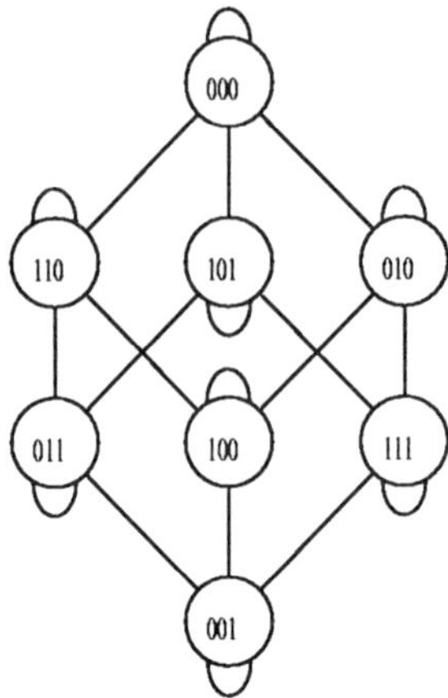

Figure 5.22 Cayley Graph Corresponding to the Example Function

The adjacency matrix A for this Cayley graph is a matrix of order $2^n \times 2^n$ with $a_{ij} = 1$ if $f(m_i \oplus m_j) = 1$ and with $a_{ij} = 0$ elsewhere. Since $m_i \oplus m_j = m_j \oplus m_i$, A is a symmetric matrix. The adjacency matrix for the example function $f = \bar{x}_1\bar{x}_3 + x_2\bar{x}_3 + x_1\bar{x}_2x_3$ is given in Equation 5.20, and a diagram representing the graph described by A is shown in Figure 5.22. Note that the figure contains self edges for each vertex because the diagonal of A contains ones. For functions that do not include the minterm $\bar{x}_1\bar{x}_2\ldots\bar{x}_n$ in f^{ON}, the adjacency matrix of the corresponding Cayley graph will contain all zeros and no self edges will be present.

$$A = \begin{bmatrix} 1 & 0 & 1 & 0 & 0 & 1 & 1 & 0 \\ 0 & 1 & 0 & 1 & 1 & 0 & 0 & 1 \\ 1 & 0 & 1 & 0 & 1 & 0 & 0 & 1 \\ 0 & 1 & 0 & 1 & 0 & 1 & 1 & 0 \\ 0 & 1 & 1 & 0 & 1 & 0 & 1 & 0 \\ 1 & 0 & 0 & 1 & 0 & 1 & 0 & 1 \\ 1 & 0 & 0 & 1 & 1 & 0 & 1 & 0 \\ 0 & 1 & 1 & 0 & 0 & 1 & 0 & 1 \end{bmatrix} \tag{5.20}$$

The spectrum of a graph is defined as the set of eigenvalues of the adjacency matrix representing it [36] (in some texts the spectrum of a graph is defined as the eigenvalues and their associated multiplicity). The theorems and proofs given in [8] demonstrate that the spectrum of the Cayley graph representing the group as is also defined in [8], which in turn represents some Boolean function f, is identical to the Walsh spectrum (in R-encoding) of the Boolean function. This result is interesting since it relates the concept of the spectrum of a graph with the Walsh spectrum of a Boolean function. The characteristic polynomial $C(\lambda)$ for the adjacency matrix given in Equation 5.20 is shown in the following.

$$C(\lambda) = \lambda^8 - 8\lambda^7 + 16\lambda^6 + 16\lambda^5 - 80\lambda^4 + 64\lambda^3$$

Solving $C(\lambda) = 0$ yields the eigenvalues, $\lambda_i \forall i = 1, \ldots, 8 = \{4, 2, 0, -2, 0, 2, 0, 2\}$. These eigenvalues are the Hadamard-Walsh spectral coefficients for f as is verified in Equation 5.21.

$$
\begin{bmatrix} 4 \\ 2 \\ 0 \\ -2 \\ 0 \\ 2 \\ 0 \\ 2 \end{bmatrix}
=
\begin{bmatrix}
1 & 1 & 1 & 1 & 1 & 1 & 1 & 1 \\
1 & -1 & 1 & -1 & 1 & -1 & 1 & -1 \\
1 & 1 & -1 & -1 & 1 & 1 & -1 & -1 \\
1 & -1 & -1 & 1 & 1 & -1 & -1 & 1 \\
1 & 1 & 1 & 1 & -1 & -1 & -1 & -1 \\
1 & -1 & 1 & -1 & -1 & 1 & -1 & 1 \\
1 & 1 & -1 & -1 & -1 & -1 & 1 & 1 \\
1 & -1 & -1 & 1 & -1 & 1 & 1 & -1
\end{bmatrix}
\begin{bmatrix} 1 \\ 0 \\ 1 \\ 0 \\ 0 \\ 1 \\ 1 \\ 0 \end{bmatrix}
\tag{5.21}
$$

Using Equation 3.2, it is possible to obtain the Hadamard-Walsh spectrum using S-encoding as expressed in the mapping relation shown in Equation 5.22. Alternatively the adjacency matrix A can be encoded using S-encoding and the resulting eigenvalues will be S-encoded Walsh spectral values.

$$
\begin{bmatrix} 4 \\ 2 \\ 0 \\ -2 \\ 0 \\ 2 \\ 0 \\ 2 \end{bmatrix} \Longrightarrow \begin{bmatrix} 0 \\ -4 \\ 0 \\ 4 \\ 0 \\ -4 \\ 0 \\ -4 \end{bmatrix} \qquad (5.22)
$$

Although the Cayley graph for the example function is strongly regular, disjoint graphs can result. In fact, for the Boolean function $f : B^n \to 0$ the corresponding Cayley graph using the group defined above results in 2^n disconnected vertices [8]. The example graph in Figure 5.22 is strongly regular due to the fact that the Walsh spectrum of the corresponding example f contains only four distinct spectral coefficients (in R-encoding). Many other interesting properties that relate this type of Cayley graphs to the Boolean function Walsh spectrum are given in [8].

Because this class of Cayley graphs and their associated adjacency matrices have exponential size in terms of n, the straight forward application of this technique is not of interest from a computational complexity point of view. However, the theoretical link demonstrating the equivalence of the spectra of Cayley graphs and Boolean function Walsh spectra can lead to insight into the interpretation and use of this set of spectral coefficients.

5.5 SPECTRA OF INCOMPLETELY-SPECIFIED FUNCTIONS

An underlying concept of this book is the use of DDs to compute the spectra of Boolean functions. The use of BDDs to represent functions necessitates the fact that all don't-care's are assigned to Boolean values. This is typically the case after synthesis has occurred. In fact, the presence of a don't-care set is desired for a synthesis tool since it represents another degree of freedom during optimization over some set of design constraints.

A don't-care set can be interpreted in differing ways in the description of a Boolean function. For this reason techniques are discussed that include the don't-care as a third logic value for a Boolean function as well as considering it as a value that may be assigned as a logic-0 or a logic-1. In the former case it is considered as an exact (albeit third) logic value while in the latter case it is considered as a value that may be arbitrarily assigned to either a logic-0 or a logic-1 value. That is, in the latter case it is assumed that the function is synthesized to realizable binary-valued digital circuitry.

An incompletely-specified function is one where certain points in the range space are unspecified and are thus commonly considered to be "don't-cares". In terms of synthesis, incompletely-specified functions can be advantageous since the don't-care values may be assigned to either a logic-1 or logic-0 value leading to an increased degree of freedom with respect to minimization of the cost function. Alternatively, in terms of the equivalence-checking problem, incompletely-specified functions may increase the difficulty of finding a solution since two completely-specified functions, each found by assigning values to the don't-cares in an incompletely-specified function, will, in general, differ for one or more assignments to the inputs. This is of significant practical interest since a circuit derived from an incompletely-specified functional specification of course implements a completely-specified function, so the issue at hand arises when determining if two circuits do indeed implement a particular function at all its specified points.

Two methods for the computation of the spectra of incompletely-specified functions are considered here. In the first, the computation of the spectra for any two of the functions $ON(f)$, $OFF(f)$ and $DC(f)$ is considered. In the second, the concept of interval arithmetic is used resulting in each spectral coefficient being computed over an interval of values rather than an exact integer.

Computation of Spectra with Assigned Don't-cares

In [90] it is pointed out that a *Multiple Valued Logic* (MVL) system may be used where each minterm in the $DC(f)$ set is assigned a unique value for some function f. Using the so-called S-encoding as is used throughout this book where the logic values $\{0, 1, -\}$ are assigned the integer values $\{-1, +1, 0\}$, respectively, yields a unique spectrum for each possible Boolean function. Alternatively, the so-called R-encoding may be used where each logic value $\{0, 1, -\}$ is assigned the integer value $\{0, +1, \frac{1}{2}\}$.

In terms of DD representations, if these spectra are desired, it is necessary to represent any two of the three possible sets $ON(f)$, $OFF(f)$ or $DC(f)$ as BDDs where the on-sets of the functions are represented with a terminal-1 value and the off-sets with a terminal-0 (or vice-versa). The spectrum of the third unspecified function is then easily obtained.

Computation of Spectra with Non-assigned Don't-cares

An alternative way to consider the spectrum of such functions is to view them as an $ON(f)$ and $OFF(f)$ set of minterms with the $DC(f)$ set as being decomposed into two disjoint subsets where each member of the subsets is arbitrarily assigned to either the $ON(f)$ or $OFF(f)$ sets. In this case, each individual spectral coefficient may be defined using interval arithmetic. That is, an individual spectral coefficient S_i may be computed as the interval value $[S_i^{LB}, S_i^{UB}]$. In this case, S_i^{LB} corresponds to the case where all minterms in $DC(f)$ are assigned to the logic-0 value, and S_i^{UB} corresponds to the case where all minterms in $DC(f)$ are assigned logic-1.

All of the techniques for the computation of spectral coefficients presented in the preceding paragraphs are easily extended to handle the case for the interval computations as well as the exact computations using the extended definitions of S- and R-encodings.

5.6 SUMMARY

The usefulness of spectral based methods in VLSI design is irrelevant if there are no efficient methods for their computation. This chapter has described several techniques for computing the spectra or subsets of spectral coefficients using DD and cube list data structures. These data structures are widely used in modern design environments; thus, the utility of these methods is justified. The traditional, Cooley-Tukey based "fast" methods, have been transformed into graph algorithms suitable for implementation over DD structures. In terms of partial spectra computations, techniques are described based on function probability computations, and mathematical matrix-algebra definitions are made based on DD structures. A method based on cube lists is presented. Finally, the relationship among the Walsh spectrum and a Cayley graph is discussed.

6

BDD MINIMIZATION

Binary Decision Diagrams (BDDs) are a powerful tool and are frequently used in many applications in VLSI CAD, such as logic synthesis and verification. However, these data structures are very sensitive to variable ordering and their resulting size often becomes intractable for practical implementation. Several techniques for variable (re-)ordering have been proposed. (For an overview see [43].)

In this chapter BDD minimization in the context of spectral techniques is discussed. After providing some background on BDD minimization techniques, a variable reordering method based on Haar spectral techniques is presented. This method is very robust and results in an efficient method to minimize the graph size [162].

The second part of this Chapter is dedicated to the study of linear transformations from both a theoretical and a practical point of view. First, it is proven for a family of Boolean functions that by linear transformations an exponential blow-up of the BDD size can be prevented [78].

Second, an exact algorithm for finding the linear transformation for which the corresponding BDD has minimal size is given [77]. An heuristic method is presented that makes use of the exact algorithm and allows one to significantly reduce the BDD sizes for large functions [76].

For both techniques experimental results are given that demonstrate the advantages of the approaches.

117

6.1   BDD VARIABLE ORDERING TECHNIQUES

Several design methods have been proposed that are based on *ordered Binary Decision Diagrams* (BDDs) [15]. The resulting circuits have very nice properties, such as good testability [3, 6] and low power dissipation characteristics [108]. For synthesis approaches based on *Pass Transistor Logic* (PTL) BDDs seem to be a good starting point. Promising results on how to transform a decision diagram to a circuit based on PTL have been reported in [9, 20, 144, 171].

One drawback of BDDs is that they are very sensitive to the variable ordering; the size of the representation may vary from linear to exponential in the number of function variables. Therefore, in the last few years several methods have been presented to determine good orderings. However, improving the optimal variable ordering starting from a given BDD representation is known to be an NP-complete problem [11].

Existing methods to determine good variable orderings can be classified into three categories:

- The first ones are initial heuristics starting from a circuit [69, 68, 113],

- the second ones are gradual improvement heuristics based on the exchange of variables in the BDD [44, 70, 93, 122, 126, 127, 135], and

- the third ones are exact methods to find an optimal ordering [48, 67, 93, 95].

Regarding the trade-off between run time of the algorithms and resulting BDD sizes, the algorithms of the second group give the best results. The BDDs are small, but the run times are moderate.

The basic operation of *Dynamic Variable Ordering* (DVO) is the exchange of adjacent variables [70, 135]. The exchange is performed very quickly since only edges must be redirected within these levels. Thus, the size is optimized without a complete reconstruction of the BDD. Only local transformations for the two levels are performed. This is due to the observation that BDDs are a canonical form. The exchange of two variables does not change the sub-BDDs of other levels.

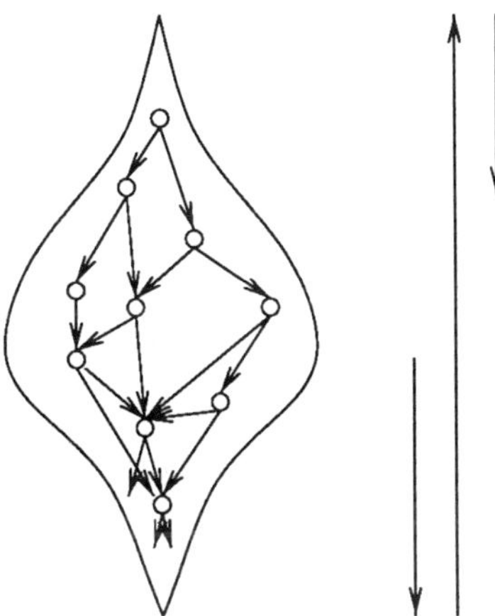

Figure 6.1   Sifting One Variable

Sifting

The most popular DVO approach is the *sifting algorithm* [135]. It successively considers all variables of a given BDD. When a variable is chosen, the goal is to find the best position of the variable, assuming that the relative order of all other variables remains the same. In the first step, the order in which the variables are considered is determined. This is done by sorting the levels according to their size with largest level first. To find the best position, the variable is moved across the whole BDD. In [135] this is done in three steps (see Figure 6.1).

1. The variable is exchanged with its successor variable until it is the last variable in the ordering.

2. The variable is exchanged with its predecessor until it is the topmost variable.

3. The variable is moved back to the closest position which has led to the minimal size of the BDD.

The algorithms can be directly generalized to other DD types including FDDs and KFDDs [42].

6.2 PROBABILITY BASED MINIMIZATION

A method for reordering the variables in a BDD to reduce the size of the data structure based on spectral techniques is presented here. A common heuristic for the variable ordering problem can be formulated as an extension of the sifting technique such that the variables with similar characteristics are grouped together [126]. This heuristic is used to formulate a technique for the reordering problem using probability based metrics. The results indicate that this technique outperforms sifting with comparable run times. Furthermore, the method is robust in that the final result is independent of the initial structure of the BDD.

6.2.1 Heuristic Description and Motivation

Many researchers have utilized heuristics that measure the relationship between primary inputs. For example, techniques based on topological closeness were among the first variable ordering methods [28, 113]. Techniques that exploit relationships between dependent variables continued to be popular particularly after it was noted that the application of sifting [135] led to orderings where symmetric variables tended to cluster together [122, 126].

The determination of symmetry among variables is not a computationally efficient process to undertake by pure manipulation of BDD structures. Therefore, an alternative method based on circuit output probabilities is formulated here to utilize symmetry indicators for the purpose of reordering.

Output Probability

The output probability of a Boolean function expresses the probability that the function will evaluate to a logic 1 value given the probability distributions of the dependent variables [128] (see also Chapter 2). When the function is fully specified and all variables are equally likely to be 0 or 1, the output probability is simply the percentage of minterms that cause the function to evaluate to logic 1. We denote the output probability of a function f as $\wp\{f\}$.

Example 6.1 $\wp\{x_1 + x_2 + x_3\} = \frac{7}{8}$

The output probability of a function may be computed with complexity that is linear in time with respect to the size of a BDD. Various algorithms have been developed that have complexity proportional to the number of vertices in the BDD [24, 103], and they are applicable to BDDs with complemented edges [119].

Since we are interested in using symmetry relations among the variables of a Boolean function, the relationship in Theorem 6.1 provides the connection between variable symmetry and output probabilities.

Theorem 6.1 $\wp\{f \oplus x_i\} = \wp\{f \oplus x_j\}$ *if* $x_i, x_j \in S$ *where* S *is the set of all independent variables that support* f *and* $x_i \leftrightarrow x_j$ *(that is,* x_i *and* x_j *are symmetric in* f*).*

Proof: From the Shannon expansion theorem we have

$$f = \overline{x}_i f_0^i + x_i f_1^i$$

Using the properties of probability theory and assuming that x_i and x_j are statistically independent, temporally and spatially uncorrelated, and equally likely to be 0 or 1 (*i.e.* $\wp\{x_i\} = \wp\{x_j\} = \frac{1}{2}$)

$$\wp\{f_{x_i}\} = \wp\{f|x_i\} = \frac{\wp\{f \cdot x_i\}}{\wp\{x_i\}} = 2\wp\{f \cdot x_i\}$$

Substituting this result into the output probability relationship for the logical XOR of a function with its ith dependent variable yields

$$\wp\{f \oplus x_i\} = \wp\{f\} + \wp\{x_i\} - 2\wp\{f \cdot x_i\}$$

$$\wp\{f \oplus x_i\} = \wp\{f\} + \wp\{x_i\} - \wp\{f_{x_i}\}$$

Clearly, $x_i \leftrightarrow x_j$ implies that $\wp\{f_{x_i}\} = \wp\{f_{x_j}\}$. Thus,

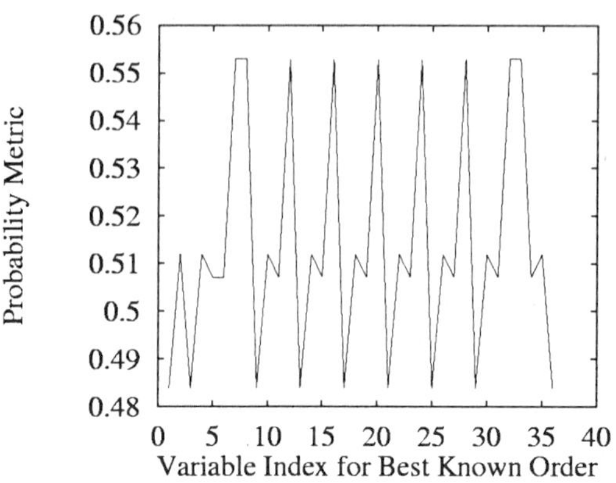

Figure 6.2 Behavior of $\wp\{f \oplus x_i\}$ versus Best Known Ordering for $c432$, Output $370gat$

$$\wp\{f\} + \wp\{x_i\} - \wp\{f_{x_i}\} = \wp\{f\} + \wp\{x_j\} - \wp\{f_{x_j}\}$$

This relationship can then be rewritten as

$$\wp\{f \oplus x_i\} = \wp\{f \oplus x_j\} \qquad \qquad \square$$

The output probabilities described here may be computed using algorithms that perform single traversals of BDDs [24, 119]. Thus, it is practical to compute the n probability values of $\wp\{f \oplus x_i\}$ for all $x_i \in S$.

Relationship of Variable Order and Output Probability

During the formulation of this technique, we generated plots of $\wp\{f \oplus x_i\}$ versus the dependent variable index i for the best known orderings of several benchmark circuits in BDD form with negative edge attributes. Due to the heuristic based on symmetric variables we expected to see probability values with similar magnitudes clustered together. In many cases, the resultant plots illustrated an unexpected trend of periodicity. This periodic trend was found in many other benchmark circuits and suggests that dependent variables with the same $\wp\{f \oplus x_i\}$ values tend to position themselves as far apart from one another as possible. Figure 6.2 illustrates the behavior using the best known ordering (see Section 6.2.2) for a circuit from the *ISCAS85* benchmark set.

In order to exploit the characteristic of periodicity, an algorithm was developed that orders the variables such that the corresponding $\wp\{f \oplus x_i\}$ values are as periodic as possible. The technique involves the following steps:

1. Choose an output about which to compute n values of $\wp\{f \oplus x_i\}$.

2. Compute all the probability values.

3. Determine histogram bin sizes and widths.

4. Form a histogram of the $\wp\{f \oplus x_i\}$ values.

5. Starting from the bin with the lowest probability value, choose a value.

6. Move to the next adjacent bin (in a circular fashion) and choose a value.

7. When all values have been removed from the bin, build a list of dependent variables in the same order in which the probability values were chosen.

8. Reorder the initial BDD according to the list just generated.

9. Perform a modified sifting routine referred to as "bin-sifting".

Figure 6.3 contains a plot of the $\wp\{f \oplus x_i\}$ values versus the index of x_i after application of the variable reordering algorithm. This data corresponds to the same output as was used in Figure 6.2, 370*gat*. Clearly, periodicity of the probability values has been enforced.

The histogram bin width has proven to be a crucial factor. Varying this parameter can affect the results. The bin width is currently calculated by the following procedure.

1. Find the two closest valued $\wp\{f \oplus x_i\}$ values that are not equal and compute their difference, δ_s.

2. Find the difference, δ_b, that corresponds to subtracting the overall smallest $\wp\{f \oplus x_i\}$ value from the largest.

3. Set BINSIZE$=\delta_b/\delta_s$.

4. If BINSIZE exceeds a threshold value, set it equal to the threshold. In the results given here a threshold of 10,000 is used.

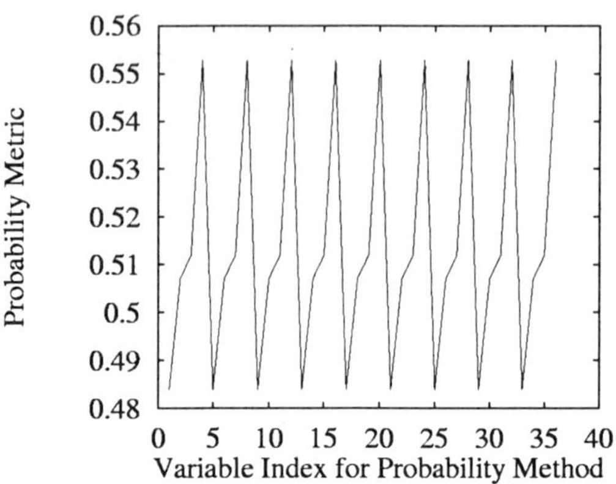

Figure 6.3   Behavior of $\wp\{f \oplus x_i\}$ versus Probability Based Ordering for $c432$, Output $370gat$

Bin-Sifting

When the order list is created by traversing the histogram, the particular variable chosen from those contained in the same histogram bin is arbitrary. Thus, the resulting intermediate BDD can be biased by the choice of which variable within a particular histogram bin is chosen. To alleviate effects caused by this bias, we invoke a modified form of the sifting algorithm referred to as "bin-sifting". In bin-sifting we use the sifting paradigm but only among variables with common $\wp\{f \oplus x_i\}$ values. That is, all variables within a certain histogram bin are sifted only amongst themselves. This preserves the periodic nature of the histogram of $\wp\{f \oplus x_i\}$ values but often leads to significantly smaller BDDs. Part of the reason this occurs is that our condition for detecting symmetry is only a necessary one. The condition is not sufficient to prove that two variables are in fact symmetrical.

The effectiveness of bin-sifting alone has been evaluated, and some results are given in the experimental results that follow. It is noted that using bin-sifting only results in BDDs that are influenced by their initial variable order since the initial order essentially "fixes" a subset of locations to which any one variable can be moved in the order list. However, bin-sifting in general seems to outperform sifting alone and may be preferable for functions that exhibit little or no variable symmetry.

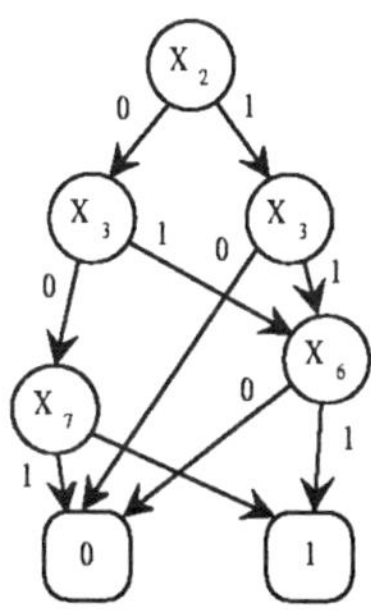

Figure 6.4 ROBDD of Example Function with Initial Variable Order

Example of the Method

To illustrate the technique described above, a small example is shown with all intermediate calculations given. We have chosen the "toy benchmark" c17 for this purpose. Although the technique was used for BDDs with CEs, here we use a BDD formulation without CEs of the c17 output labeled **23gat** for simplicity.

Initially, the BDD exists with a variable order $x_2 \prec x_3 \prec x_6 \prec x_7$ as shown graphically in Figure 6.4. The first step of the process is to compute the output probabilities, $\wp\{f \oplus x_i\}\forall i$. This step requires n traversals of the BDD. The computed output probabilities for the functions $f \oplus x_i$ are: $\wp\{f \oplus x_2\} = 0.6875$, $\wp\{f \oplus x_3\} = 0.3125$, $\wp\{f \oplus x_6\} = 0.6875$ and $\wp\{f \oplus x_7\} = 0.3125$.

A histogram is formed using the probability metrics to guide which variables are grouped together. Figure 6.5 contains the histogram. A new variable order is obtained by visiting the histogram bins and removing a variable in a circular fashion, thus ensuring periodicity of the $\wp\{f \oplus x_i\}$ values. The new order obtained from the histogram is x_7, x_6, x_3, x_2. A BDD is formed by reordering the initial structure according to this new order. The intermediate BDD is shown graphically in Figure 6.6.

The last step of the reordering technique involves the application of the bin-sifting routine. Since the example contains two histogram bins, the bin-sifting approach allows for exchanging the locations of $\{x_3, x_7\}$ and $\{x_6, x_2\}$ only. We note that if all probability metrics had been equal, the bin-sifting approach degenerates into the standard sifting algorithm. The resulting variable order for the example function is found to be x_3, x_6, x_7, x_2. The final BDD is shown

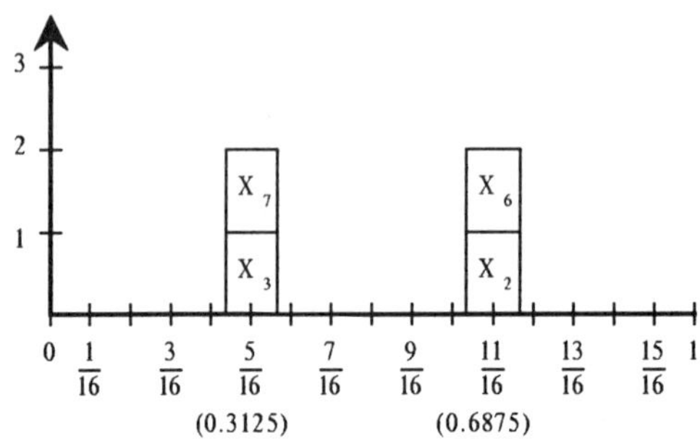

Figure 6.5 Histogram of Example BDD Variables Based on Probability Metrics

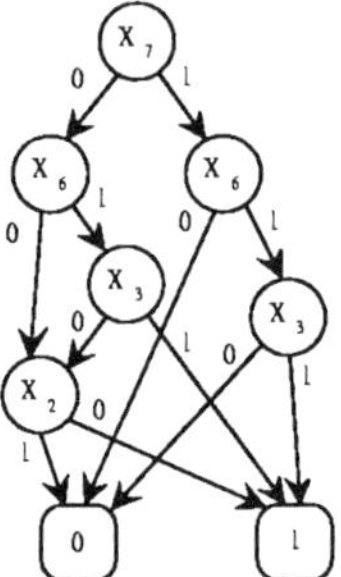

Figure 6.6 ROBDD of Example Function with Intermediate Variable Order

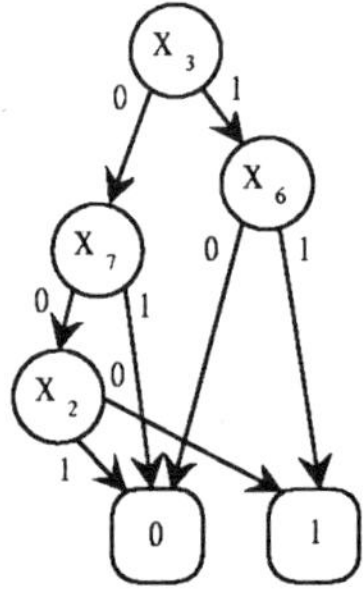

Figure 6.7 ROBDD of Example Function with Final Variable Order

in Figure 6.7. In summary, we note that the size of the example BDD was initially 5 vertices and resulted in 4 vertices for the final result.

6.2.2   Experimental Results

Variable orderings have been computed for several benchmark circuits using the techniques described above. These results have been compared to sifting alone which yields reordering improvements in a small amount of CPU time. As compared to the best known orderings, which are computed using techniques such as those based on *Simulated Annealing* (SA) and *Evolutionary Algorithms* (EAs); these results yield very good minimized BDDs.

The first set of experimental data compares the results of the probability-based heuristic to that of sifting. Table 6.1 contains reordering results for some of the `pla` benchmarks. Many of the *pla* circuits yield non-interesting results in that the size of the resulting BDD is the same regardless of the order, thus the small subset in Table 6.1. The initial ordering was arbitrarily set to the order in which the variables appear in the *pla* files. In all experiments performed using the *pla* benchmarks, the size of the resulting BDD was always less than or equal to the original size. In some cases the reduction was up to 20%.

Table 6.2 contains the data resulting from the technique, sifting alone, and bin-sifting alone. Results for benchmarks c2670 and c7552 were not included since attempts to convert these into a BDD with a random initial variable order resulted in exceeding the maximum BDD vertex limit in the program. Circuit c6288 is also not included since it represents a multiplier that is known to not have any size less than exponential in the number of variables.

In all cases the probability method performed as well or better than sifting or bin-sifting alone in terms of resultant BDD size. The increase in CPU time is less than an order of magnitude as compared to sifting in most cases. The results using bin-sifting alone illustrate that it generally performs as well or better than sifting, but at an increase in CPU time due primarily to the large size of the initial BDD structure.

The results in Tables 6.1 and 6.2 also contain a column for total CPU time. This is the time that was required for generating the probabilities, creating and traversing the histogram, reordering according to histogram traversal, and performing bin-sifting. The time was obtained using a Sun SPARCstation 20 with 240 MB of RAM, a single 100 MHz processor and running under Solaris 2.5. It must be noted that these times assume that it is known prior to execution which output will be used for computing the $\wp\{f \oplus x_i\}$ values.

Table 6.1 Experimental Results for the Reordering Technique Using *PLA* Benchmarks

circuit		sifting		prob. method	
Name	Org. Size	Size	Time	Size	Time
5xp1	74	51	< 1	42	< 1
alu4	1197	800	< 1	639	1
bw	108	103	< 1	99	< 1
duke2	973	394	< 1	339	1
misex1	41	37	< 1	37	< 1
misex2	136	89	< 1	84	< 1
misex3	1301	704	< 1	683	5
sao2	155	94	< 1	87	< 1
misex3c	828	504	< 1	472	1
clip	226	87	< 1	87	< 1
e64	1441	231	1	231	2
apex1	28336	1356	3	1333	58
apex4	928	895	< 1	889	1
apex5	2679	1130	< 1	1130	1

Currently an automated technique for determining the BDD output about which the output probabilities should be computed has not been implemented. The results given above are randomly selected outputs from the benchmark functions. It is noted that there is typically not a wide variance in the resulting BDD sizes based on the selected output. As proof of this phenomena, Table 6.3 contains the maximum, minimum, average and standard deviation of the sizes of BDDs computed using several different outputs for some of the *ISCAS85* benchmark circuits. The standard deviation values indicate the extent of how the output selection affects the resultant BDD size.

Table **6.2** Experimental Results for the Reordering Technique Using *ISCAS*85 Benchmarks

benchmark		sift		bin-sift		prob. meth.	
Circuit	Org. Size	Size	Time	Size	Time	Size	Time
c432	31172	20891	28	20891	216	1119	13
c499	53866	38196	84	36234	156	34432	245
c880	10348	5089	9	4603	20	4603	26
c1355	53866	38186	88	36234	159	34432	280
c1908	13934	8175	11	7762	23	6170	48
c3540	72858	41918	73	37644	1421	38096	257
c5315	747662	4078	225	-	-	3447	4108

Table **6.3** BDD Size Statistics Versus Output About Which Probabilities are Computed

circuit name	number of outputs	max size	min size	avg size	stan. dev.
c432	6	1636	1119	1340	173
c499	32	36250	34432	35784	781
c880	12	31990	4603	7276	3197
c1355	32	36250	34432	35784	781
c1908	25	8174	6170	7607	2185

6.3 LINEAR TRANSFORMATIONS

Several extensions of the basic BDD structure have been proposed in the last few years (see Chapter 4). A very promising extension is based on spectral techniques, since the BDD data structure itself remains unchanged, and only an encoding of the inputs is carried out. Instead of transforming the function table, the input variables are transformed by *linear transformations*. In contrast to spectral methods that operate with $2^n \times 2^n$-matrices, for linear transformations $n \times n$-matrices are sufficient. An efficient heuristic for BDD minimization under linear transformations has been presented in [115], which is called *linear sifting*. The sifting algorithm has been extended by an effective linear operator which combines neighboring variables of the BDD. The algorithm runs very fast, since the basic operation is (very similar to variable reordering [70]) a local operation. By this, it is possible to obtain much smaller BDD sizes than with sifting alone. Although it is not noted in [70, 115], linear transformation of BDDs is simply the graphical counterpart of type *(iv)* spectral translation as described in Chapter 3.

In the following, linear transformations from a theoretical and practical point of view are studied. First, an exponential difference in the BDD sizes for linear transformations is proven. The exponential difference results in the observation that there exists a set of functions for which the BDD size is exponential; however, there is a linear transformation under which the BDD has linear size. Second, we describe an exact algorithm for BDD minimization under linear transformations, that is, an algorithm that finds a linear transformation for which the size of the BDD is minimal with respect to the number of nodes. As variable reordering is a subset of linear transformations, the approach is more powerful than reordering alone. However, we make use of reordering techniques, since they allow us to prune large parts of the search space without loss of exactness. Then we present a window optimization algorithm for heuristic BDD minimization. It can be seen as an extension of the window permutation algorithm [93] for dynamic BDD minimization to the case of linear transformations. The window permutation algorithm exactly optimizes small windows of neighboring variables, and these windows are shifted over the BDD. This is also the main idea for the linear case, but instead of using permutations only, linear transformations can also be used which makes the algorithm more powerful. Experimental results are given to demonstrate the advantage of the approach.

Some useful basic definitions are given in the following.

Definition 6.1 *A re-encoding of a set of input variables $X = \{x_1, \ldots, x_n\}$ is an automorphism $\tau : B^n \to B^n$, which is a mapping for which the inverse mapping exists. The re-encoded input variables are denoted by $\tau(x_1, \ldots, x_n)$.*

In synthesis, this generally means the realization of the re-encoding and the transformed function (see Figure 6.8).

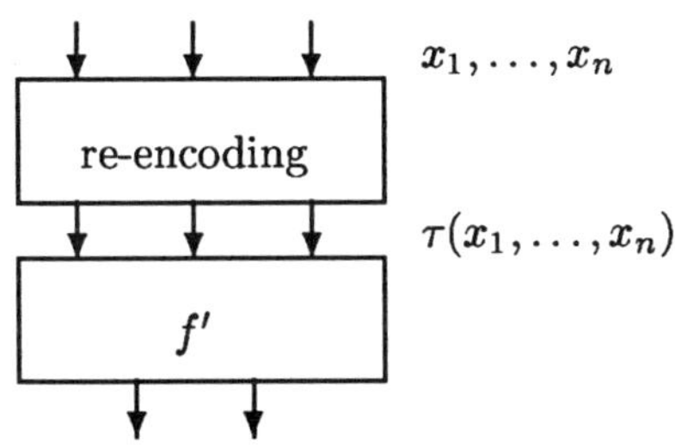

Figure 6.8 Realization in Synthesis

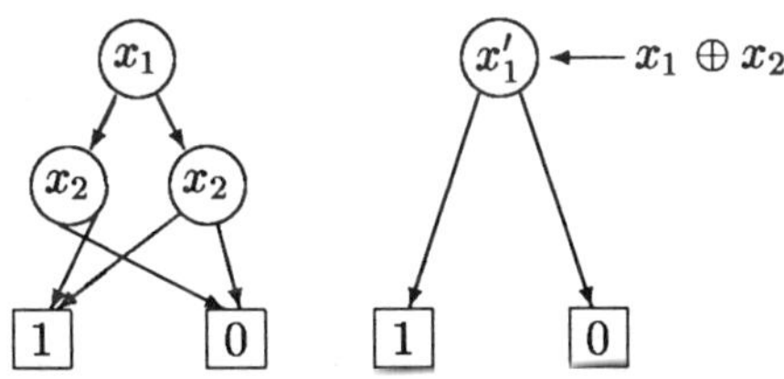

Figure 6.9 BDDs of the Function $f = \overline{x}_1 x_2 + x_1 \overline{x}_2$

Example 6.2 *For function $f = \overline{x}_1 x_2 + x_1 \overline{x}_2$, the transformation $\tau(x_1, x_2) = (x_1 \oplus x_2, x_2)$ is an automorphism. The BDDs for f with and without transformation τ are given in Figure 6.9.*

As described above, we restrict ourselves to linear transformations which will be defined in the following.

Definition 6.2 *For $i \neq j$, an elementary transformation $\sigma_{ij} : B^n \to B^n$ is an automorphism of type*

$$\sigma_{ij}(x_1, \ldots, x_n) = (x_1, \ldots, x_{i-1}, x_i \oplus x_j, x_{i+1}, \ldots, x_n).$$

An elementary transformation can be written as an $n \times n$-matrix over the Galois field (B , $\oplus$, $\cdot$):

$$
\sigma_{ij}(x_1,\ldots,x_n) = \begin{array}{c} \\ i \end{array}\begin{pmatrix} 1 & & & & 0 \\ & \ddots & & 1 \\ & & \ddots & & \\ 0 & & & & 1 \end{pmatrix} \cdot \begin{pmatrix} x_1 \\ \vdots \\ \\ \vdots \\ x_n \end{pmatrix}
$$

(All non-given matrix elements are 0.)

Linear transformations are those automorphisms obtained by a sequence of elementary transformations [115]. This is accomplished in a manner that is analogous to the generation of regular matrices using elementary matrices.

Definition 6.3 *A linear transformation is a linear re-encoding of the set of input variables, i.e. a re-encoding which can be represented by a regular $n \times n$-matrix over the Galois field* (B , $\oplus$, $\cdot$).

As shown in [116] the number of possible linear transformations is

$$
\prod_{i=0}^{n-1}(2^n - 2^i),
$$

which is much larger than $n!$, the number of possible variable orderings, but much smaller than $2^n!$, the number of all possible automorphisms. Similar estimations hold for the space requirements: permutations require $O(n)$ space, linear transformations $O(n^2)$, and general automorphisms $O(2^n)$. Therefore, linear transformations are a compromise, as they can be represented efficiently and enlarge the search space significantly which gives more room for optimizations.

In the following we denote a permutation of the variables with π. If $x_i = \pi(k)$ for a variable x_i, then x_i is the kth element of the variable ordering π, *i.e.* x_i is in the kth *level* of the BDD. For a set of variables $I \subseteq X$, let $\Pi(I)$ be the set of permutations π that map the variables from I to I. (We also write $\pi(I) = I$.)

For linear transformations τ_1 and τ_2, let $\tau_2 \circ \tau_1$ denote the concatenation of τ_1 and τ_2, *i.e.* the linear transformation $(\tau_2 \circ \tau_1)(x_1,\ldots,x_n) = \tau_2(\tau_1(x_1,\ldots,x_n))$.

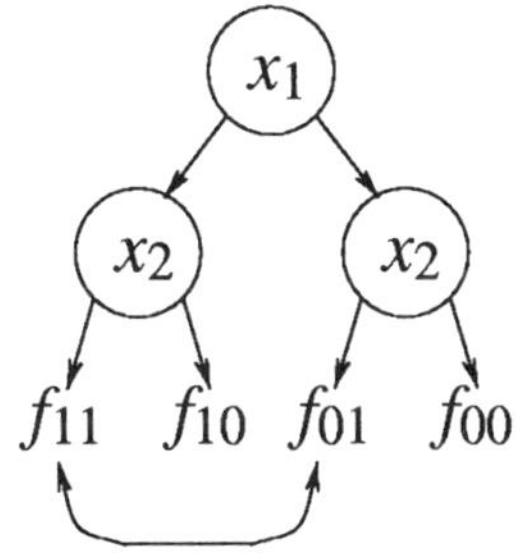

Figure 6.10   Performing a Linear Transformation

Identifying linear transformations as regular matrices, this can be seen as a multiplication of two regular matrices.

The number of nodes in a BDD for function f and linear transformation τ is given by $\#nodes(f,\tau)$. The nodes in level k marked with variable x_i under linear transformation τ are given by $\#nodes_{x_i}(f,\pi)$.

6.3.1   Linear Transformations on BDDs

Performing a linear transformation on a BDD can be done efficiently using a technique that is similar to *level exchanges* in the case of variable reordering. It will be described here how a linear combination of two adjacent levels (*i.e.* setting an elementary transformation) can be realized on BDDs.

A Shannon decomposition for a node representing function f with respect to the variables x_1 and x_2 leads to

$$f = x_1 x_2 f_{11} + x_1 \overline{x}_2 f_{10} + \overline{x}_1 x_2 f_{01} + \overline{x}_1 \overline{x}_2 f_{00}.$$

Setting the elementary transformation σ_{12} means replacing x_1 by $x_1 \oplus x_2$ in the formula above, thus

$$f = x_1 x_2 f_{01} + x_1 \overline{x}_2 f_{10} + \overline{x}_1 x_2 f_{11} + \overline{x}_1 \overline{x}_2 f_{00}.$$

This transformation is a local operation on BDDs, since only two outgoing edges have to be exchanged, namely f_{11} and f_{01} (see Figure 6.10).

If complement edges are used f_{01} can be complemented, whereas f_{11} must not be complemented. Thus, the exchange operation can be non-local. However, locality can be preserved using the complemented XOR $\overline{\oplus}$ instead of $\oplus$:

$$f = x_1 x_2 f_{11} + x_1 \overline{x}_2 f_{00} + \overline{x}_1 x_2 f_{01} + \overline{x}_1 \overline{x}_2 f_{10},$$

i.e. exchanging f_{10} and f_{00}.

In the following the theory will be described using the uncomplemented XOR, but the complemented XOR will be taken in the experiments. It is easy to see that the theory holds for both cases.

6.3.2 Exponential Gap

We now present a theoretical result which demonstrates the power of linear transformations to BDD minimization: there exists a family of Boolean functions for which BDDs are exponential in size under each variable ordering, but there are linearly transformed BDDs of linear size. For the proof an argument of lower bound computation from [16] is needed and is briefly reviewed.

Let $f : B^n \to B$ be a Boolean function and let (L, R) be a partition of X. Then, any input assignment $x : X \to B$ can be split into a *left input assignment* $l : L \to B$ and a *right input assignment* $r : R \to B$. We use the notation $x = lr$. A set $\mathbf{F}_{(L,R)}$ of left input assignments is called a *fooling set* for the partition (L, R) if, and only if, for each two distinct $l, l' \in \mathbf{F}_{(L,R)}$ there is a right input assignment r with $f(lr) \neq f(l'r)$. Fixing a real parameter $\omega \in (0, 1)$, a partition (L, R) is called a *balanced partition* if, and only if,

$$\lfloor \omega \cdot n \rfloor \leq |L| \leq \lceil \omega \cdot n \rceil.$$

Lemma 6.1 *[16] Assume that f has a fooling set $\mathbf{F}_{(L,R)}$ of size c for each balanced partition (L, R) with respect to some ω. Then the BDD for f has at least c nodes[1].*

To simplify the main proof, we prove a rather simple lemma first.

Definition 6.4 *Consider an $n \times n$-square of points (see Figure 6.11). A line is either a horizontal row or a vertical column of the square. A line is* covered *if, and only if, all its points are covered, and a line is* active *if, and only if, at least one point of the line is covered and at least one point of the line is uncovered.*

---

[1] This number includes constant nodes.

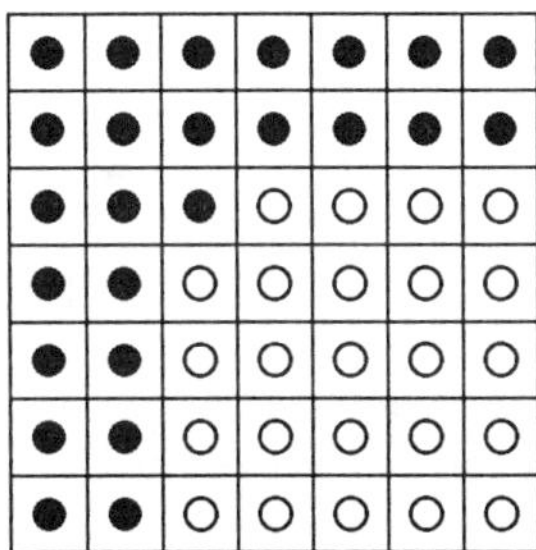

Figure 6.11 A 7 × 7 Square of Points (see Lemma 6.2)

Lemma 6.2 *Let* $\lfloor 0.5n^2 \rfloor \leq k \leq \lceil 0.5n^2 \rceil$. *If k points of an $n \times n$-square of points are covered, then there are at least n active lines in the square.*

Proof: Let k points of an $n \times n$-square be covered, and let k be within the given range. We distinguish the following four cases.

Case 1: There are only horizontal covered lines. Then all n vertical lines are active.

Case 2: There are only vertical covered lines. This case is analogous to Case 1.

Case 3: There are both horizontal and vertical covered lines. Then all of the $2 \cdot n$ lines of the square have a covered point. Furthermore, we can assume the covered lines to be at the left and at the top of the square, since exchanging lines does not change the number of active lines. An example for this situation in case of $n = 7$ is given in Figure 6.11. Since about half of the points are covered, there is an uncovered rectangle of about half of the points on the lower right of the square (the uncovered area can even be larger). Furthermore, there are at least as many active horizontal lines as the rectangle's vertical length and at least as many active vertical lines as the rectangle's horizontal length. As the sum of horizontal and vertical lines is minimal for a given constant area in case of a square, we can assume that the uncovered rectangle is a square.

Since by assumption there are at least $0.5n^2 - 1$ uncovered points, the uncovered square has a size of at least

$$\lfloor \sqrt{0.5n^2 - 1} \rfloor \geq \sqrt{0.5n^2 - 1} - 1.$$

Since for $n \geq 5$, it holds

$$\frac{1}{2}n^2 - 1 \geq \frac{4}{9}n^2,$$

the size is at least

$$\sqrt{\frac{4}{9}n^2} - 1 = \frac{2}{3}n - 1 \geq \frac{1}{2}n,$$

whereas the last inequality holds for $n \geq 6$. Thus, for $n \geq 6$, the square has size at least $\frac{1}{2}n$. As the number of active lines is double the size, there are at least n active lines. (Asymptotically, there are even $\sqrt{2} \cdot n$ such lines. Furthermore, a straightforward computation shows that the same property holds for $2 \leq n < 6$.)

Case 4: There are neither horizontal nor vertical covered lines. A computation similar to the previous case shows again the minimum number of n active lines. $\qquad\square$

Theorem 6.2 *There exists a family of Boolean functions $f : B^{n^2} \to B$ such that any BDD representing the function has size $\Omega(2^n)$, but there is a linear transformation such that the transformed function can be represented by a BDD of size $2 \cdot n + 1$.*

Proof: Consider the function $f : B^{n^2} \to B$ given by

$$f(x_1, \ldots, x_{n^2}) = \sum_{i=0}^{n-1} \bigoplus_{j=1}^{n} x_{n \cdot i + j} + \sum_{i=1}^{n-1} \bigoplus_{j=0}^{n-1} x_{n \cdot j + i}.$$

For $n = 3$, a BDD representing the function is given in Figure 6.12.

The function can be seen as a disjunction of $2 \cdot n - 1$ XOR chains. Assuming the n^2 variables in an $n \times n$-matrix, these XOR chains are the XOR of horizontal or vertical lines:

$$\begin{pmatrix} x_1 & x_2 & \cdots & x_n \\ x_{n+1} & x_{n+2} & & x_{2n} \\ \vdots & & \ddots & \vdots \\ x_{n(n-1)+1} & x_{n(n-1)+2} & \cdots & x_{n^2} \end{pmatrix}$$

Note that the right vertical XOR chain consisting of the variables $x_n, x_{2n}, \ldots, x_{n^2}$ is missing.

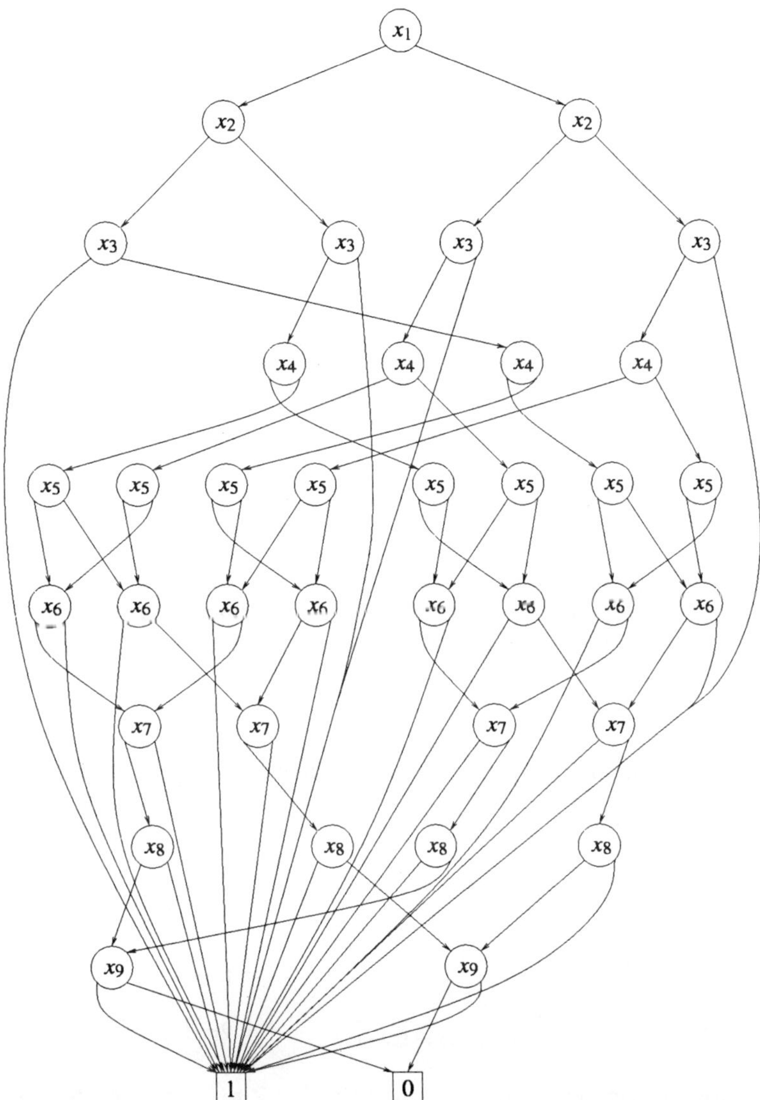

Figure 6.12 A BDD for the Function in the Proof of Theorem 6.2 ($n = 3$)

The proof for the lower bound of the resulting BDD size is based on Lemma 6.1. Let $\omega = 0.5$ and (L, R) be an arbitrary balanced input partition. Then, by Lemma 6.2 there are at least n active XOR chains. As the vertical line of

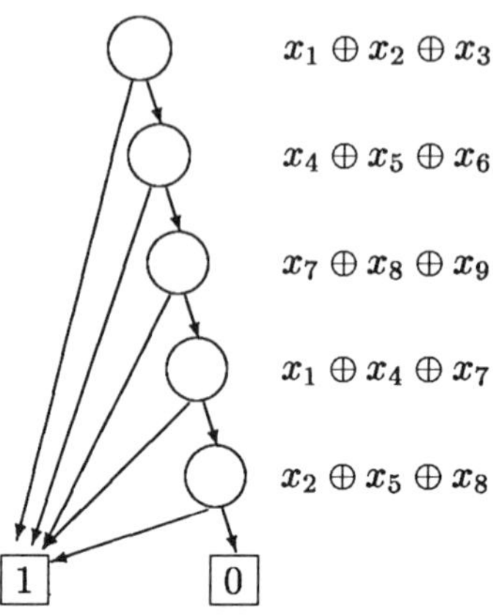

Figure 6.13 The Transformed BDD for $n = 3$.

the matrix is missing and thus cannot be covered, one of the claimed n XOR chains which the lemma claims might be missing. Since satisfiability of one of these XOR chains is independent of another, the fooling set $\mathbf{F}_{(L,R)}$ has at least size 2^{n-1}. Thus, by Lemma 6.1, the BDD representing f has at least $\Omega(2^n)$ nodes.

Now we have to show that there is a linear transformation such that the BDD under that transformation has $2 \cdot n + 1$ nodes. A straightforward computation shows that for the transformation $\tau(x) = (\tau_1(x), \ldots, \tau_{n^2}(x))$ given by

$$
\tau_i(x_1, \ldots, x_{n^2}) = \begin{cases} \displaystyle\bigoplus_{j=1}^{n} x_{n \cdot (i-1)+j} & i = 1, \ldots, n \\[2em] \displaystyle\bigoplus_{j=0}^{n-1} x_{n \cdot j + i - n} & i = n+1, \ldots, 2n-1 \\[2em] x_{i-n+2+\lfloor \frac{i-2n}{n-1} \rfloor} & i = 2n, \ldots, n^2, \end{cases}
$$

the resulting function is the disjunction of the first $2 \cdot n - 1$ transformed variables and thus the BDD needs $2 \cdot n + 1$ nodes. (Note that the lines $2n, \ldots, n^2$ of the transformation ensure that the transformation is an automorphism.) This number includes the constant nodes (see Figure 6.13 for $n = 3$). $\square$

Thus, there exist functions that can not be handled efficiently using BDDs, while linear transformed BDDs allow a very compact representation. An experimental evaluation of the function is given in Section 6.3.5.

6.3.3 Exact Minimization

In the last section the theoretical quality of linear transformations has been studied. In this section we describe a method to exactly determine the best linear transformation. In other words, we consider the following problem:

> *How can we determine an optimal linear transformation, i.e. a linear transformation such that the number of BDD nodes is minimal?*

This problem can be seen as an extension to the problem of finding the optimal variable ordering for a given Boolean function, since the set of permutations is a subset to the set of linear transformations. Since techniques from optimal variable ordering are used in the algorithm, these techniques are shortly reviewed in the following.

Optimal Variable Ordering

In [67] an exact algorithm has been presented where the number of variable orderings which have to be considered could be reduced significantly in comparison with the trivial idea of constructing BDDs for all $n!$ variable orderings. The main idea behind that algorithm is given in the following lemma.

Lemma 6.3 *[67] Let $f : \mathbf{B}^n \to \mathbf{B}^m$, $I \subseteq X$, $k = |I|$, and $x_i \in I$. Then there exists a constant c such that $\#nodes_{x_i}(f, \pi) = c$ for each $\pi \in \Pi(I)$ with $\pi(k) = x_i$.*

More informally, the lemma states that the number of nodes in a level is constant if the ordering of the variables above that level or below that level is changed (see Figure 6.14).

The lemma can be used for variable reordering as follows: for $I \subseteq X$, let

$$min\_cost_I = \min_{\pi \in \Pi(I)} \sum_{i=1}^{|I|} \#nodes_{x_i}(f, \pi),$$

and let π_I denote some permutation leading to that minimum. Assume for a fixed $I \subseteq X$ with $|I| = k$ that we know $min\_cost_{I'}$ for all $I' \subseteq I$ with $|I'| = k-1$. Then we get

$$min\_cost_I = \min_{x_i \in I}(min\_cost_{I-\{x_i\}} + \#nodes_{x_i}),$$

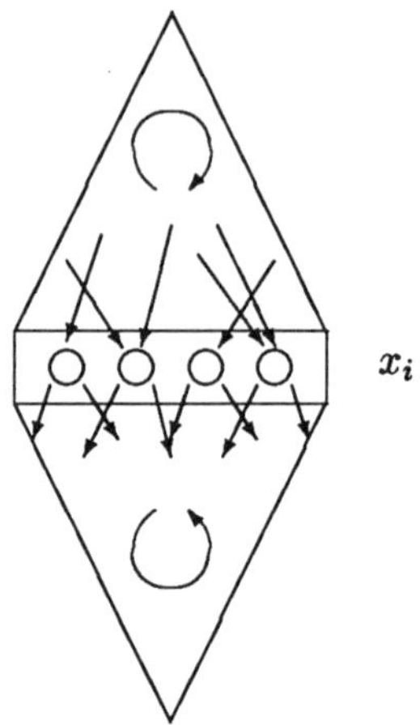

Figure 6.14   Graphical Illustration of Friedman's Lemma (Lemma 6.3)

with $\#nodes_{x_i}$ being the number of nodes in level k under some permutation π with $\pi(I - \{x_i\}) = I - \{x_i\}$ and $\pi(k) = x_i$. So the optimal ordering can be computed iteratively by computing min_cost_I for each k-element subset I for increasing k's, until $k = n$.

Remark 6.1 For increasing k the considered level k moves down. Likewise, one can start at the top level and consider the levels upwards. As $\Pi(I) = \Pi(X - I)$, Lemma 6.3 can be used the same way in that case. (For a detailed discussion see [48, 49]).

Linear Transformations

In this section some theory about linear transformations is given that is necessary for the understanding of the exact algorithm. The following lemma is the key to the approach, as it allows the decomposition of general linear transformations into a permutation and a "normalized" linear transformation. By this, the algorithm can also make use of techniques from the case of optimal variable ordering, especially Lemma 6.3.

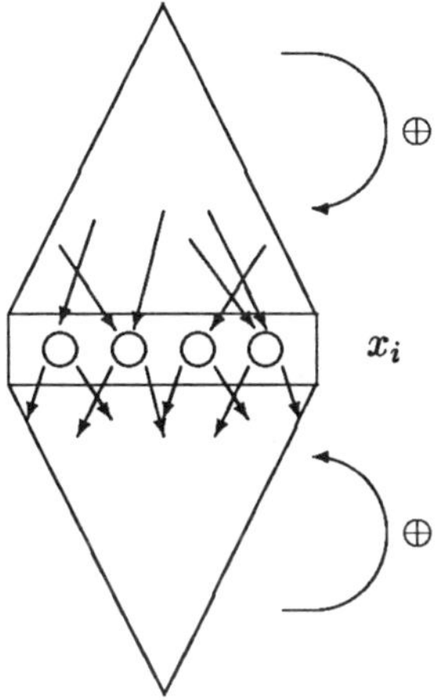

Figure 6.15   Graphical Illustration of Theorem 6.3.

Lemma 6.4 *For any linear transformation τ, there exists a permutation π and a linear transformation ϱ such that $\tau = \pi \circ \varrho$ and for matrix $R = (r_{ij})_{i,j}$ of ϱ it holds that $r_{i,i} = 1$ for $1 \leq i, j \leq n$. We call ϱ a normalized* linear transformation.

Proof: Let $T = (t_{ij})_{i,j}$ be the $n \times n$-matrix of τ over $\mathbf{B}$. As T is regular, there exists p_1 such that $t_{p_1,1} = 1$ (otherwise, the first column of T would be the vector 0, in contradiction to T being a regular matrix). In other words there exists a permutation $\pi^{(1)}$ such that $T = \pi^{(1)} \cdot R^{(1)}$ for some $R^{(1)} = (r_{ij}^{(1)})_{i,j}$ with $r_{11}^{(1)} = 1$.

As $R^{(1)}$ is regular, by induction we get $\pi^{(2)}, \ldots, \pi^{(n)}$ and $R^{(n)} = (r_{ij}^{(n)})_{i,j}$ such that $r_{ii}^{(n)} = 1$ for all $1 \leq i \leq n$. For $\pi := \pi^{(1)} \cdot \ldots \cdot \pi^{(n)}$ and $R := R^{(n)}$ the postulated conditions hold. $\square$

As a consequence, instead of constructing BDDs for all linear transformations, we need only construct the BDDs for all permutations and for all normalized linear transformations. The following theorem can be seen as an extension of Lemma 6.3 to normalized linear transformations.

Theorem 6.3 *Let $f : \mathbf{B}^n \to \mathbf{B}^m$. Then there exists a constant c such that $\#nodes_{x_i}(f, \varrho) = c$ for each normalized linear transformation ϱ with corresponding matrix $(r_{ij})_{i,j}$ and*

1. *$r_{ii} = 1$ for all $1 \le i \le n$,*
2. *$r_{ij} = 0$ for all $i < k$ and $j \ge k$,*
3. *$r_{ij} = 0$ for all $i \ge k$ and $j < k$,*
4. *$r_{ij} = 0$ for all $i > k$ and $j \le k$,*

i.e. a matrix of type:

$$
\begin{pmatrix}
1 & & * & 0 & \cdots\cdots & 0 \\
 & \ddots & & \vdots & & \vdots \\
* & & 1 & 0 & \cdots\cdots & 0 \\
\hline
0 \cdots & 0 & & 1 & * \cdots & * \\
\vdots & & 0 & 0 & 1 & * \\
\vdots & & & \vdots & & \ddots \\
0 \cdots & \cdots & 0 & * & & 1
\end{pmatrix}
$$

Proof: A similar technique can be used as was employed in the proof of Lemma 6.3. Without any loss of generality, the variable ordering is $(x_1, x_2, \ldots, x_n)$, *i.e.* $\pi(k) = x_k$ for all $1 \le k \le n$. Elementary transformations within the lower part of the BDD, *i.e.* within the levels $k + 1, \ldots, n$, do not affect the number of nodes in level k, as each such node corresponds to a cofactor of the BDD's functions which depends on the variable x_k. The number of such cofactors is independent of the encoding of $x_{k+1}, \ldots, x_n$.

Additionally, applying elementary transformations σ_{ki} with $i > k$ does not change the number of nodes in level k, as their number is independent of their encoding. Cofactors which do not depend on x_k also do not depend on $x_k \oplus x_i$.

Elementary transformations within the upper part of the BDD, *i.e.* within the levels $1, \ldots, k - 1$, do not affect the number of nodes in level k, as the set of cofactors which have to be represented by nodes in the lower part is independent of the encoding of the upper variables. In particular, the number of cofactors which depend on x_k does not change. $\qquad\Box$

This means that we can apply elementary transformations σ_{ij} with $i < k$ and $j < k$ or with $i > k$ and $j \geq k$ without changing the number of nodes in level k.

Furthermore, we have to prove that by this method we traverse the whole search space.

Theorem 6.4 *Any normalized regular $n \times n$-matrix R can be written in the form*

$$R = U_n \cdot D_n \cdot U_{n-1} \cdot D_{n-1} \cdot \ldots \cdot U_1 \cdot D_1$$

with

$$U_k = \bordermatrix{ & & k & k+1 & \cr & 1 & & & & \cr & & \ddots & & & \cr k & & & 1 & u^{(k)}_{k+1} \cdots u^{(k)}_n \cr & & & & \ddots & \cr & & & & & 1 } \quad \text{and} \quad D_k = \bordermatrix{ & & & k & \cr & 1 & & & \cr & & \ddots & & \cr k & & & 1 & \cr k+1 & & & d^{(k)}_{k+1} & \ddots \cr & & & \vdots & \ddots \cr & & & d^{(k)}_n & & 1 }.$$

(All non-given matrix elements are 0.)

Proof: For $n = 1$, $R = (1)$ is the only regular matrix, so for $U_1 = (1)$ and $D_1 = (1)$ the equation holds.

For $n > 1$, let $R = (r_{ij})_{i,j}$. Let M' be the matrix obtained from R by deleting the first row and the first column, *i.e.* $M' = (r_{ij})_{2 \leq i,j \leq n}$. By induction we know about the existence of $U'_n, D'_n, \ldots, U'_2, D'_2$ such that $M' = U'_n \cdot D'_n \cdot \ldots \cdot U'_2 \cdot D'_2$. We can enlarge all these matrices by one leading column and row, *i.e.* we get

$$M = \begin{pmatrix} 1 & 0 & \cdots & 0 \\ 0 & r_{22} & \cdots & r_{2n} \\ \vdots & \vdots & \ddots & \vdots \\ 0 & r_{n2} & \cdots & r_{nn} \end{pmatrix}.$$

Evidently it holds $M = U_n \cdot D_n \cdot \ldots \cdot U_2 \cdot D_2$. We set $u_i^{(1)} := r_{1,i}$ for $i = 2, \ldots, n$, so we have:

$$
M \cdot U_1 = \left(\begin{array}{c|ccc}
1 & r_{12} & \cdots & r_{1n} \\
0 & r_{22} & \cdots & r_{2n} \\
\vdots & \vdots & \ddots & \vdots \\
0 & r_{n2} & \cdots & r_{nn}
\end{array} \right)
$$

As $M \cdot U_1$ is a regular matrix, we can define

$$
\left(\begin{array}{c}
d_1^{(1)} \\
\vdots \\
d_n^{(1)}
\end{array} \right) := (M \cdot U_1)^{-1} \cdot \left(\begin{array}{c}
r_{11} \\
\vdots \\
r_{n1}
\end{array} \right)
$$

and we finally have $M \cdot U_1 \cdot D_1 =$

$$
\left(\begin{array}{c|ccc}
1 & r_{12} & \cdots & r_{1n} \\
0 & r_{22} & \cdots & r_{2n} \\
\vdots & \vdots & \ddots & \vdots \\
0 & r_{n2} & \cdots & r_{nn}
\end{array} \right) \cdot \left(\begin{array}{c|ccc}
d_1^{(1)} & 0 & \cdots & 0 \\
d_2^{(1)} & 1 & & 0 \\
\vdots & & \ddots & \\
d_n^{(1)} & 0 & & 1
\end{array} \right) = R
$$

as D_1 is regular, it must hold $d_1^{(1)} = 1$. $\square$

It can easily be seen that all matrices of the intermediate steps are normalized. So we can prune the search space whenever we get an "un-normalized" matrix. We do not mention this further, as it would complicate the algorithm unnecessarily.

Finding the Optimal Linear Transformation

Algorithms for finding the optimal variable ordering are based on iteratively considering all levels of the BDD [48, 67]. Considering one level means calculating the minimum number of nodes for this level using the calculation of the step before. In the case of reordering, this has to be done for all k-element subsets $I \subseteq X$ with increasing k. Furthermore, in our more complex case we additionally have to consider linear transformations, as shown in the previous section. In this regard our algorithm can be seen as an extension to reordering algorithms.

```
compute_optimal_linear_transformation (f) {
    init table with linear transformation;
    for all BDD levels k (top→down) {
        next_table := ∅;
        for each entry t in table {
            set t's permutation and linear transformation;
            prune if lower bound is large enough;
            for each variable xᵢ in lower levels {
                for each linear transformation of xᵢ and lower levels {
                    c := actual cost;
                    key := key of new permutation / linear transformation;
                    if (key ∉ next_table or c < next_table[key])
                        add to next_table;
                }
            }
        }
    }
    set best permutation and linear transformation;
}
```

Figure 6.16   Sketch of the exact algorithm

A sketch of the exact algorithm is given in Figure 6.16. For increasing k, we consider all subsets I of size k and all linear transformations ϱ based on elementary transformations σ_{ij} with $i < k < j$ or $j < k < i$. For each such combination we store the number of nodes in levels 1 to k (*min_cost*), the best permutation, and the best linear transformation for these levels in a table. We use (I, ϱ) as a hash key for our tables.

The table is initialized with $I = \emptyset$, *min_cost* $= 0$, some initial ordering π, and linear transformation ϱ. For any level k (for increasing k) we consider all entries in *table*. Without pruning this would be every subset $I \subseteq X$ of size k and every linear transformation ϱ based on elementary transformations σ_{ij} with $i < k$ and $j \geq k$ or with $i \geq k$ and $j < k$.

Then the remaining variables x_i which are not in $X - I$ are shifted to level k. Notice that this does not change the number of nodes in levels $1, \ldots, k - 1$. For each variable x_i in level k, all linear transformations of lower variables x_j with x_i and all linear transformations of x_i with lower variables x_j are set (see

Theorem 6.4). We store the result in *next_table* either if no such element is in it or if the number of nodes in levels $1, \ldots, k$ is less than before.

In the last iteration, *i.e.* $k = n$, the minimum BDD size over all variable orderings and normalized linear transformations is computed.

Remark 6.2 *When applied in synthesis, the linear transformation also has to be realized which can be done using chains of XOR gates. In general, the number of additional gates used is small compared to the gain counted in number of BDD nodes. However, when interested in the exact result this number is also of importance.*

Lower Bound Technique

The lower bound is the smallest number of nodes possible with a given permutation and linear transformation for levels $1, \ldots, k$. Obviously, we can not determine the optimal number within reasonable time, so we have to give an estimation for that number which can be computed efficiently.

Lower bounds for the size of BDDs have been proven in [16] using lower bound techniques from VLSI design (see Lemma 6.1). These techniques are also applied in an automated way [48] giving good estimations for the lower bound in case of "pure" reordering. We do not repeat the exact method here, but it can be easily seen that due to Theorem 6.3 the same technique can also be used in the case of linear transformations. (For more details see [48].)

6.3.4 Heuristic Methods

Because of the huge search space, the exact method is only applicable to small functions. To minimize larger functions, it is therefore necessary to use heuristic methods to overcome this limitation. In this section we describe two heuristics which can be used to find good results within a reasonable amount of time.

Linear Sifting

An effective heuristic approach for BDD minimization using linear transformations has been presented in [115]. It can be seen as an extension of sifting [135]

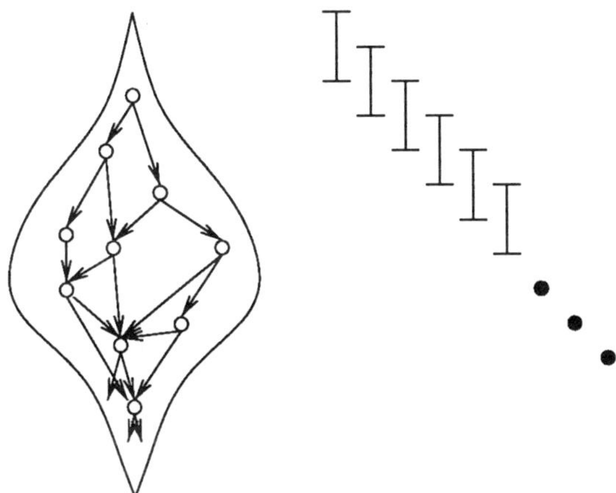

Figure 6.17 Considered Windows for the Window Optimization Algorithm

by a linear operator: instead of only considering level exchanges, linear sifting also considers the effect of the linear combination of adjacent levels.

As in sifting, linear sifting is based on the consideration of all of the variables. For each variable, the order of all other variables remains fixed and the variable is moved up and down using level exchanges and linear combinations, then it is moved to the position that leads to the smallest size of the BDD (see Figure 6.1).

Using linear sifting, significant reductions of the BDD sizes are possible. The algorithm is also applicable to large functions. Following [115] the costs for the realization of the transformation are not considered here but can easily be incorporated.

Window Optimization

A more time consuming algorithm is a generalization of the window permutation algorithm [93, 135] for BDD minimization under permutations to the case of linear transformations. Here, small windows of neighboring variables are minimized exactly using the algorithm described above. The considered windows are moved over the BDD as long as there is an improvement (see Figure 6.17).

More precisely, for all possible window positions, a record is kept that indicates whether the window position needs to be considered (in the beginning

```
1     window_optimization(f, size) {
2         linear_sifting(f);
3         for (k = 2; k ≤ size; k + +) {
4             do {
5                 mark positions 1 to n − size + 1;
6                 while there is a marked position p {
7                     optimize window p, .., p + size − 1;
8                     if (improvement) {
9                         mark positions p − 1 and p + 1;
10                    }
11                }
12                sifting(f);
13            } while (improvement by sifting);
14        }
15  }
```

Figure 6.18   The Window Optimization Algorithm

all positions are selected). If a window has been optimized and the BDD size could be reduced, the neighboring positions are marked again. This process is repeated as long as there are positions to consider. Then, sifting can be called. (Note that linear sifting could result in higher costs, as the cost for the linear transformation is not counted. Therefore, pure sifting is used here.) If sifting improves the solution, all window positions are marked to be considered again. Otherwise, the window size is increased until a maximum limit is reached. It is furthermore possible to call linear sifting at the beginning to get a good initial solution. A sketch of the algorithm is given in Figure 6.18.

Due to the large runtime of the exact method, tiny windows have to be chosen. However, the size of the window has a large effect on the quality of the results (see Section 6.3.5).

If the algorithm is modified to start with the maximum window size, *i.e.* the loop in Line 3 is omitted, the resulting BDD sizes become much worse. The time that is needed to optimize small windows is almost negligible compared to the larger windows. By first optimizing small windows, the total BDD size can be reduced which then speeds up the computation for the larger windows. Furthermore, the processing of small window sizes seem to have a positive global effect.

Table 6.4 BDD Sizes for the Function of Section 6.3.2

n	in	BDD	linear
3	9	37	6
4	16	149	8
5	25	461	10
6	36	1357	12
7	49	3821	14
8	64	11261	16
9	81	29429	18
10	100	69101	20
11	121	172013	22

6.3.5 Experimental Results

In this section we describe experimental results that have been carried out on a *SUN Ultra 1*. For all our experiments we used an upper memory limit of 100 MBytes. All times are given in CPU seconds. The algorithm has been implemented using the CUDD package [148].

In the first series of experiments we studied the actual size of the BDD and the linearly transformed BDD for the "constructed" function given in Section 6.3.2. The results which are given in Table 6.4 clearly demonstrate the exponential growth for the BDD sizes (in terms of the square root of the number of variables) but the linear growth for the linearly transformed BDD.

In a second series of experiments we examined the exact algorithm. First, all XOR gates that are needed for the realization of the linear transformations are counted as zero. Results for all LGSynth93 benchmarks with less than 7 inputs are given in Table 6.5. In column *in/out* the number of inputs and outputs is given, respectively. In column *reorder* the minimum number of BDD nodes under all variable orderings is given. Results for linear sifting [115] are given in column *linear sifting* measured for a fixed set of parameters as given in [148]. The minimum number of nodes for all linear transformations is given in column *exact.* Columns *time* and *MB* refer to runtime in seconds and memory usage

Table 6.5 Results of the Exact Algorithm

circuit	in/out	reorder	linear sifting	exact	time	MB
cm42a	4/ 10	20	20	20	1.2	1
cm82a	5/ 3	12	8	7	11.7	1
c17	5/ 2	7	9	7	21.7	1
decod	5/ 16	32	32	32	17.0	1
majority	5/ 1	8	6	6	19.6	1
rd53-hdl	5/ 3	17	15	10	10.3	1
cm138a	6/ 8	18	18	18	591.4	7
bcd.div3	4/ 4	14	15	12	1.1	1
dc1	4/ 7	22	23	21	1.2	1
dekoder	4/ 7	19	19	17	0.9	1
newcwp	4/ 5	9	9	8	0.3	1
wim	4/ 7	20	21	18	0.9	1
bw	5/ 28	101	108	95	59.9	1
xor5	5/ 1	6	4	2	0.3	1
newbyte	5/ 8	17	17	17	6.3	1
p82	5/ 14	56	58	50	39.4	1
bench	6/ 8	89	89	78	10127.1	59
fout	6/ 10	119	118	108	12179.8	59
m1	6/ 12	44	44	42	1272.1	13
newapla2	6/ 7	17	17	17	164.4	3
pope.rom	6/ 48	203	222	197	7496.8	20
sqr6	6/ 12	63	63	61	7858.7	59
sum		913	935	843		

in MB of our exact algorithm, respectively. In line *sum*, the sum of nodes for
the columns is given.

While on average linear sifting does not reduce the sizes compared to exact variable ordering, the use of linear transformations can reduce the sizes, sometimes even significantly, for benchmark examples, *i.e.* up to 40% (see *e.g.* rd53-hdl and p82).

Furthermore, we compared the minimum number of nodes when giving a "penalty" p for each elementary linear transformation, *i.e.* each entry in the linear transformation costs as much as p additional BDD nodes. The results are given in Table 6.6. In columns $p = 0$, $p = 1$ and $p = 2$, the sum of BDD size and p times the number of elementary linear transformations is given, respectively[2].

Even with penalties (which is more realistic for synthesis) linear transformations can improve the costs. The higher the penalties the less linear transformations have to be considered during the minimization process. Thus, our algorithm becomes even more efficient with respect to run time in this case, and larger functions can be handled within the space limitations.

In the following the window optimization algorithm is examined. The influence of the window size is shown in Table 6.7. For each circuit three window sizes are considered. For these experiments we minimized the number of BDD nodes and the number of XOR nodes needed to carry out the transformation in parallel. (Notice that when using linear transformed BDDs in synthesis also the cost of the transformation function has to be considered.) Column *cost* gives the sum of BDD nodes and XOR nodes using penalty $p = 1$. As can be seen, a larger window size often results in better BDD sizes (see *e.g.* c1355), but can also require much larger run times (see *e.g.* c3540).

A comparison of the sifting algorithm, the linear sifting algorithm, and our window optimization algorithm is given in Table 6.8. For sifting and linear sifting we used the implementation of the CUDD package [148]. Since linear sifting does not optimize the cost needed for the XOR realization, we explicitly give both the BDD node count (column *nodes*) and the cost for the realization of the linear transformation (column *XOR*). For the window optimization algorithm we chose a window size of 3, by which the run times remain moderate. The results obtained by linear sifting can often be further improved, sometimes even significantly (see *e.g.* c1355), while the run time overhead is small. Especially, in applications such as synthesis a reduction of more than a factor of two tremendously improves the result. The new approach is also more robust (see the functionally equivalent benchmarks c499 and c1355).

Finally, we compared the best-ever values from [148] for pure reordering to the best-ever values of linear sifting [115] and to window optimization presented

---

[2]Sharing XOR gates between the transformations has not been considered.

above (see Table 6.9). Both linear sifting and window optimization algorithms take very good variable ordering as initial ordering. Window optimization clearly out performs the results of both other approaches. In some cases an improvement of nearly a factor of three is possible (see c1908).

In summary, the experiments clearly demonstrate how powerful linear transformations are for BDD minimization. For many functions large reductions are possible. For functions with up to 7 variables we are able to find the exact optimum. For larger functions the minimization heuristic based on window optimization clearly out performs the best known heuristic. Furthermore, the new approach is also more robust.

6.4 SUMMARY

In this chapter we described approaches based on spectral techniques for minimization of BDDs.

A variable reordering method for BDDs has been presented and described. Experimental results indicate that run times are roughly equivalent to those of sifting; however, smaller BDDs generally result. The method differs significantly from those previously published in the way that function properties are used. This technique yields a resultant BDD that is independent of the initial structure and is thus very robust.

Linear transformations have been shown to be an effective tool for BDD based synthesis and verification. There are functions where an exponential reduction of the BDD size is possible, which means that by using linear transformations an exponential blow-up can be avoided. Furthermore, we gave an exact algorithm for finding the optimal linear transformation. We also presented a heuristic algorithm based on window optimization. Since the window's size can also be controlled by the user, the run time of the algorithm can be adapted dynamically. The experimental results indicated that the window optimization algorithm could often significantly reduce the BDD size (with small runtime overhead). Using the technique, improvements of up to a factor of 3 over the best known results were obtained.

Table 6.6 Costs for Different Penalties

circuit	in/out	reorder	$p=0$ cost	$p=0$ time	$p=1$ cost	$p=1$ time	$p=2$ cost	$p=2$ time
cm42a	4/ 10	20	20	1.2	20	1.0	20	0.6
cm82a	5/ 3	12	7	11.7	10	13.6	12	11.8
c17	5/ 2	7	7	21.7	7	12.3	7	4.7
decod	5/ 16	32	32	17.0	32	14.7	32	10.8
majority	5/ 1	8	6	19.6	8	19.5	8	8.8
rd53-hdl	5/ 3	17	10	10.3	15	16.8	16	11.8
cm138a	6/ 8	18	18	591.4	18	46.3	18	17.3
5xp1-hdl	7/ 10	42	—	—	41	30739.9	42	17714.8
s27	7/ 4	10	10	9912.4	10	3312.0	10	736.0
z4ml	7/ 4	17	10	15593.5	14	17773.7	17	15138.1
z4ml-hdl	7/ 4	17	10	13590.2	14	12800.8	17	11198.5
bcd.div3	4/ 4	14	12	1.1	14	1.0	14	0.6
dc1	4/ 7	22	21	1.2	22	1.0	22	0.7
dekoder	4/ 7	19	17	0.9	19	1.0	19	0.7
newcwp	4/ 5	9	8	0.3	9	0.3	9	0.3
wim	4/ 7	20	18	0.9	19	0.7	20	0.5
bw	5/ 28	101	95	59.9	97	62.7	99	65.6
rd53	5/ 3	17	10	10.1	15	16.4	16	11.9
xor5	5/ 1	6	2	0.3	6	10.7	6	2.8
newbyte	5/ 8	17	17	6.3	17	5.0	17	2.9
p82	5/ 14	56	50	39.4	53	40.0	55	42.3
bench	6/ 8	89	78	10127.1	85	10177.7	88	10477.2
fout	6/ 10	119	108	12179.8	113	12044.0	115	12025.1
m1	6/ 12	44	42	1272.2	44	1394.3	44	1450.1
newapla2	6/ 7	17	17	164.4	17	67.6	17	27.4
pope.rom	6/ 48	203	197	7496.8	200	7170.9	201	6544.3
sqr6	6/ 12	63	61	7858.7	63	6192.2	63	3894.4

Table 6.7 Different Window Sizes

circuit	in	out	size=2		size=3		size=4	
			cost	time	cost	time	cost	time
alu4	14	8	355	1.0	353	2.9	328	71.4
s1196	32	32	718	1.5	613	13.4	613	129.6
c1908	33	25	1182	11.1	1081	41.1	1034	303.6
c432	36	7	1210	1.2	1210	8.6	1210	336.9
c499	41	32	876	59.2	871	58.1	733	563.9
c1355	41	32	887	237.9	702	296.9	602	1429.3
c3540	50	22	23832	119.1	23830	350.8	23376	20932.6
c880	60	26	8449	28.2	8432	76.9	8410	4556.3
dalu	75	16	782	5.2	782	13.9	781	272.5
s1423	91	79	1466	10.9	1456	24.6	1456	423.5
c5315	178	123	1553	75.7	1549	102.1	1512	1269.8
c7552	207	108	4778	216.3	4711	379.2	4676	2403.8
c2670	233	140	2600	134.6	2500	401.2	2472	2162.6
i10	257	224	53676	3720.5	53319	4577.8	53319	31403.2

Table 6.8 Sifting vs. Linear Sifting vs. Window Optimization

circuit	in	out	Sifting		Linear Sifting			Window Optimization		
			nodes	time	nodes	XOR	time	nodes	XOR	time
alu4	14	8	429	0.3	411	10	0.5	346	7	2.9
s1196	32	32	598	0.4	706	13	1.0	605	8	13.4
c1908	33	25	6253	4.3	1416	88	9.1	983	98	41.1
c432	36	7	1210	0.3	1210	0	0.6	1210	0	8.6
c499	41	32	26408	23.3	3803	292	43.2	590	281	58.1
c1355	41	32	29562	24.0	12224	161	218.4	458	244	296.9
c3540	50	22	23828	52.2	23813	22	101.6	23811	19	350.8
c880	60	26	8409	7.6	8436	29	21.4	8417	15	76.9
dalu	75	16	829	1.2	780	4	3.2	778	4	13.9
s1423	91	79	1767	3.0	1464	29	5.4	1444	12	24.6
c5315	178	123	1778	4.5	1476	136	39.7	1421	128	102.1
c7552	207	108	7906	23.7	4657	289	134.5	4473	238	379.2
c2670	233	140	3985	8.6	2747	218	22.1	2322	178	401.2
i10	257	224	67791	234.9	53887	120	1905.3	53236	83	4577.8

Table 6.9 Comparison of Best-Ever Values

circuit	in	out	Reordering	Linear Sifting	Window Optimization
alu4	14	8	350	350	305
s1196	32	32	598	599	597
c1908	33	25	5526	1822	696
c432	36	7	1064	1209	1064
c499	41	32	25866	520	356
c1355	41	32	25866	484	371
c3540	50	22	23828	23823	23358
c880	60	26	4053	4080	4044
dalu	75	16	689	594	588
s1423	91	79	1796	1477	1365
c5315	178	123	1719	1808	1353
c7552	207	108	2212	1921	1001
c2670	233	140	1774	998	839
i10	257	224	20660	19295	19446

7

LOGIC SYNTHESIS

Logic synthesis is generally considered to be a mature technology and is utilized to some extent in the design of all modern integrated circuit devices. The most popular approach for logic synthesis is the use of methods in the Boolean domain. However, many logic synthesis operations can be carried out in the spectral domain, and some tasks such as Boolean matching and function decomposition are actually easier to perform in the spectral rather than the Boolean domain for many cases. This chapter will present some spectral-based techniques for performing logic synthesis.

7.1 SPECTRAL TRANSLATION

Section 3.5.2 described the property of spectral translation and outlined how it is useful for function classification. Spectral translation operations result in the spectral coefficients of a function being arranged in a different order but not changing in value. In Section 6.3 the spectral translation operation (iv) called "input translation" in Table 3.1, is used to reduce the size of BDDs since the corresponding DD operation (linear sifting) can be efficiently implemented.

Synthesis methods based on spectral translation have been proposed in [54, 55, 82, 92, 97, 164]. Particularly interesting is the approach in [82] since BDDs are used as the underlying data representation in contrast to the other methods. The use of spectral translation in logic synthesis involves partitioning the function to be synthesized into a linear block in the form of a set of parity functions followed by a general logic block. Figure 7.1 shows a diagram representing the partition of an example function $f(X)$ into an n-input, n-output linear

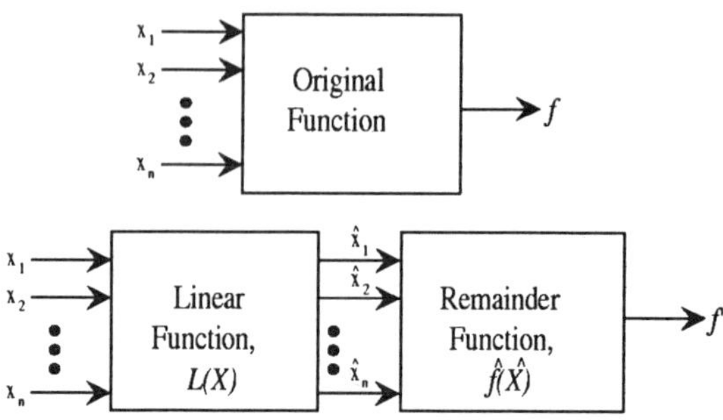

Figure 7.1   Function Realized with Spectral Translation

function $L(x)$, and an n-input single output remainder function $\hat{f}(\hat{X})$ where $X = (x_1, x_2, \ldots, x_n)$ and $\hat{X} = (\hat{x}_1, \hat{x}_2, \ldots, \hat{x}_n)$. Each $\hat{x}_i$ is a parity function depending on x_i and zero or more other variables, x_j.

The objective of the synthesis process is to determine some linear mapping $L(X)$ such that the resulting $\hat{f}(\hat{X})$ is simplified. In [82] the function complexity denoted by $C(f)$ is used to determine the linear mapping. In this case, function complexity is equal to the number of variable assignments X_i and X_j that result in a different output for a function $(f(X_i) \neq f(X_j))$ such that X_i and X_j differ by only a single bit (that is, X_i and X_j have a Hamming distance of one) [92]. This measure is useful for spectral translation since it is invariant over NPN-functions (the set of operations that result from spectral translations of type (i), (ii) and (iii) in Table 3.5.2). The value of $C(f)$ may be computed directly from the Walsh spectrum of a function and ranges in value from 0 for $f = 0$ or $f = 1$ to a value of $n2^n$ when f is a parity function. In general, the lower the value of $C(f)$, the more efficient an AND/OR representation of f becomes. An iterative algorithm based on the repeated computation of complexity measures using the DD based technique described in [34] for the spectral computations can yield good results in terms of area and delay minimization as reported in [82].

In general, the spectral translation method for logic synthesis can be used regardless of the method used to determine the $L(X)$. Example 7.1 shows a small function represented in BDD form and a corresponding linear mapping resulting in a smaller diagram.

Example 7.1 *Consider the function* $f = \overline{w}\,\overline{y}z + \overline{x}\,\overline{y}z + xyz + wy\overline{z}$ *whose BDD is shown in Figure 7.2.*

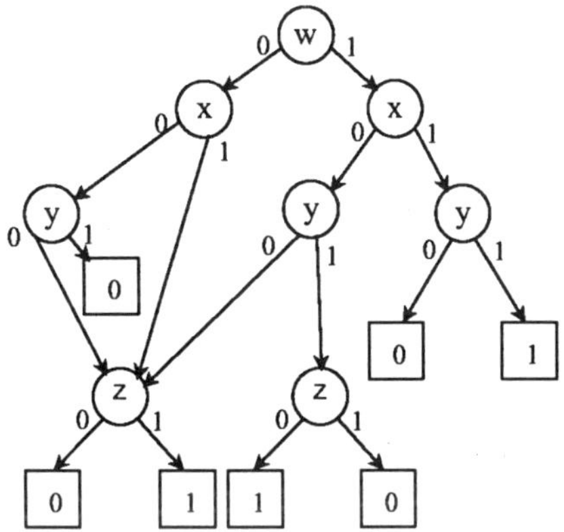

Figure 7.2 Example Function BDD Before Spectral Translation

Using the following linear mapping

$$\hat{w} = w \oplus z \qquad (7.1)$$
$$\hat{x} = x \oplus y \oplus z$$
$$\hat{y} = y \oplus z$$
$$\hat{z} = z$$

results in the simpler remainder function $\hat{f} = \hat{x}\hat{z} + \hat{w}\hat{y}$ whose BDD is shown in Figure 7.3.

7.2 MAXIMUM CORRELATION ITERATIVE APPROACH

The design and development of digital systems is becoming increasingly complex and automated. With the evolution of greater amounts of circuitry per single chip comes a need for greater optimization and efficiency in the design process. Spectral-based design techniques can offer the advantage of providing for the design of circuits in a methodical way and are, therefore, good candidates for incorporation into an automated environment. In the past there have

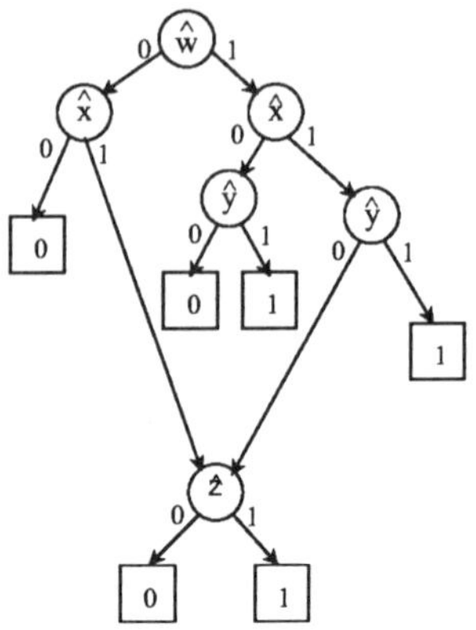

Figure 7.3   Remainder Function BDD After Spectral Translation

been a number of results that use various spectral techniques for both binary-valued logic [92, 110, 111, 155, 160, 161] as well as multi-valued and threshold logic [37, 54, 130].

The approach described here is characterized by the fact that the synthesized circuit is allowed to be composed of any of several types of sub-functions such as AND, OR, combinations of AND/OR/Invert (AOI), and more complicated cells, whereas, most previous approaches focused on the realization of logic circuitry with a single type of circuit as the basic building block. In Chapter 3 and other sources, it is shown that the Walsh spectral transform provides Boolean function output correlation measures with respect to various combinations of input variables added together via modulo-2 arithmetic [89, 92]. It is the output correlation characteristic that is exploited in the development of this method.

Examples of synthesized circuits are given that were chosen to illustrate various characteristics of this approach. In the examples given the design method is shown to be applicable to both two-level and multi-level circuits. The use of specific design parameters such as the desired gate type and maximum number of inputs per gate are considered with methods to include these types of constraints into the synthesis technique described. Computer program implementation considerations such as storage and computation requirements are discussed along with methods to minimize the usage of these resources. In addition, the basis for a computer implementation of this procedure is described, and its convergence is proven.

Typically, designers generate Boolean expressions at some point in the logic circuit design process regardless of the technique used. In fact, many logic design techniques are considered to be accomplished upon the realization of a Boolean function that is minimized in some sense. In reality, circuits are built from a schematic diagram or some other form of circuit description such as a netlist in some standard CAD tool file format. This synthesis technique does not require an intermediate Boolean function to be minimized. Furthermore, the XOR gate is fully exploited as a potential candidate for incorporation into the design in addition to the AND and OR gates. Most design techniques require symbolic manipulation of Boolean functions to achieve inclusion of the XOR gate. Indeed, it has been stated that algebraic techniques in general are not well suited for circuit realization using XOR circuits. The algebraic methods presented here are rigorously developed and are very well suited for XOR type circuits.

The rest of this section is organized as follows. The subsection entitled "Spectrum Interpretations and Properties" briefly summarizes some of the characteristics of the Walsh spectrum that this technique is based upon. Since this method utilizes a subset of the spectrum, any of the partial spectral computation methods described in Chapter 5 may be used. Several lemmas are also given that give the mathematical basis for the technique. Next, the synthesis algorithm is given and explained followed by two examples. The first example shows how a two-level logic circuit using multi-input gates may be realized using the synthesis method. The second example illustrates how the synthesis technique can be applied to a circuit that is restricted to being composed of two-input OR gates as much as possible. The second example also shows how this technique is applied for the realization of multi-level circuits. After the examples are given a discussion of the implementation issues for the synthesis methodology is included.

7.2.1   Spectrum Interpretations and Properties

In this section the following terms are used. The *Output Vector of a Boolean Function* refers to a concatenation of all possible outputs of the function using the *S*-encoding. *S*-encoding is used to allow the zero-valued function outputs to accumulate in each spectral coefficient. *Constituent Functions* denoted by f_c are Boolean functions whose output vectors are used as the rows of the transformation matrix.

Each spectral coefficient in the transformed output vector provides a measure of correlation between a particular constituent function and the original function $f(x)$. This is the underlying principle that is exploited in the synthesis procedure described here.

In general, any set of constituent functions may be used for this technique as long as they form a functionally complete set of operators for Boolean algebra in order to assure convergence. Although a transform must be constructed from an orthogonal set of basis functions for the spectrum and original function to form a unique pair, this technique does not rely on transform pair uniqueness. For this reason it is possible to use transform matrices that are not necessarily orthogonal or square. The transformation matrix must be full rank, however, for convergence of this algorithm to be guaranteed. It has been shown that a transform using the Exclusive-OR operator as the primitive operation for the constituent functions results in the Walsh transform matrix [54]. In the examples transforms are constructed using the OR and AND operators in addition to the XOR to form the constituent functions.

The notion of an output vector of a Boolean function was described earlier. A qualitative discussion of the derivation of other transform matrices extends this notion such that the rows of the transformation matrix are viewed as output vectors of the constituent functions. The constituent functions are generally simple functions using only a single operator (although they need not be). For instance, if it is desired to compute the correlation between a constituent function, $x * y$, and a function to be transformed, one row of the transformation matrix would consist of the output vector of the constituent function. Hence, the spectral coefficient resulting from the dot product of this row and the function output vector provides a measure of correlation between the overall function and the constituent function, $x * y$. In fact, a measure of correlation with any arbitrary constituent function may be computed in this manner. Each correlation measure (or spectral coefficient) contains the information of the exact number of matching outputs between the constituent function and the transformed function for a given common set of inputs. Before giving a precise relationship between the spectral coefficients and the number of matching outputs, the following notation is defined.

1. n - the number of input variables of a Boolean function

2. N - all possible combinations of the input variables of the binary-valued Boolean function, $N = 2^n$

3. N_m - the number of matches between the outputs of the constituent function and the function to be synthesized for the same input values

4. N_{mm} - the number of mismatches between the outputs of the constituent function and the function to be synthesized

5. $S_f[f_c(x)]$ - the spectral coefficient of $f(x)$ that corresponds to the constituent function, $f_c(x)$

6. $S_f(1)$ - the spectral coefficient of $f(x)$ corresponding to $f_c(x) = 1$

7. $\{f_c(x)\}$ - a set of constituent functions

8. $\underline{S_f}[\{f_c(x)\}]$ - a vector of spectral values computed over the set of constituent functions, $\{f_c(x)\}$

Some of the properties of spectral coefficients are presented in the following and are used in the development of the synthesis technique.

Lemma 7.1 *For a given function $f(x)$ and a given constituent function $f_c(x)$ the resulting spectral coefficient is given by*

$$S_f[f_c(x)] = 2^n - 2N_{mm} = 2N_m - 2^n \qquad (7.2)$$

Where x is an n-order vector of binary valued inputs to both $f(x)$ and $f_c(x)$.

Proof: The maximum possible absolute value of a spectral coefficient occurs when a row of the matrix is equal to the function output vector or when each component of the vector is the negative of the corresponding entry in the transform matrix row. Hence, the maximum possible absolute value of the spectral coefficient is $|S_f[f_c(x)]| = 2^n$ indicating 100% positive or negative correlation between $f(x)$ and $f_c(x)$. Indeed, in this case either $f(x) = f_c(x)$ or $f(x) = \overline{f}_c(x)$. Each mismatch present in the function output vector and the corresponding matrix row entry always produces a product value of -1. Therefore, N_{mm} mismatches result in a negative partial sum of $-N_{mm}$. The only other possibility is a match which is the complement of mismatches and always produces a product value of +1. Since the spectral coefficient for $f(x)$ and $f_c(x)$ is the difference between the number of matches, N_m, and the number of mismatches, N_{mm}

$$\begin{aligned} S_f[f_c(x)] &= N_m - N_{mm} \\ &= N_m - [2^n - N_m] \\ &= 2N_m - 2^n \end{aligned}$$

Likewise, substituting N_{mm}

$$\begin{aligned} S_f[f_c(x)] &= N_m - N_{mm} \\ &= [2^n - N_{mm}] - N_{mm} \\ &= 2^n - 2N_{mm} \end{aligned}$$

Hence, $S_f[f_c(x)] = 2^n - 2N_{mm} = 2N_m - 2^n$. $\qquad\square$

Lemma 7.2 *The following property of spectral coefficients holds*

$$S_f[f_c(x)] = -S_f[\overline{f}_c(x)] \tag{7.3}$$

Proof: Let the number of mismatches between the inverse of the constituent function $\overline{f}_c(x)$ be denoted by N'_{mm} and the corresponding matches denoted by N'_m.

Thus, $N'_m = N_{mm}$. Using this fact and the results from Lemma 7.1

$$\begin{aligned} S_f[f_c(x)] &= 2^n - 2N_{mm} \\ &= 2^n - 2N'_m \\ &= -(2N'_m - 2^n) \\ &= -S_f[\overline{f}_c(x)] \end{aligned}$$

$\qquad\square$

These results are used to develop logic circuit synthesis methods for combinational circuits. The following section describes the iterative synthesis method

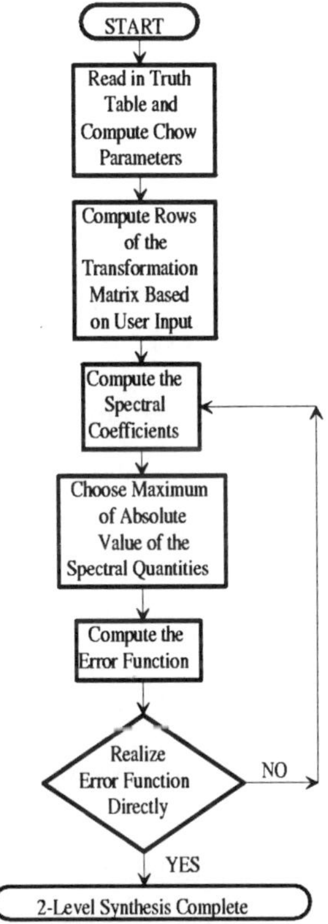

Figure 7.4 Flowchart of Two-Level Synthesis Methodology

that lends easily to automation and allows for the resulting circuit realization to be customized with respect to the maximum number of inputs per logic gate and the type of logic gate to be used.

7.2.2 Maximum Correlation Synthesis Algorithm

The synthesis method described here is dependent on the use of error functions in an iterative fashion. Figure 7.4 illustrates the overall structure of the synthesis methodology.

User supplied input consists of a description of the truth table of the function $f(x)$ to be synthesized, and optionally, the maximum number of inputs per gate N_{inp} and preferences of the types of gates $\{G_t\}$ to be used. In a practical implementation the description of $f(x)$ would likely be in the form of a cube list or a DD. To simplify the description of this method, it is assumed that the output vector of the function to be synthesized is used. This allows the small examples to be given using vector-matrix product calculations for the spectra. The two optional parameters N_{inp} and the set $\{G_t\}$ are used to determine the set of constituent functions $\{f_c(x)\}$ to be used in the formation of the transformation matrix. For the sake of generality, we will first assume that the optional parameters are not supplied; however, they are used in the second synthesis example.

The following list gives a detailed description of each synthesis step for the process depicted in Figure 7.4.

1. Convert the input truth table to 1's and -1's using the S-encoding (the integer, +1 to denote a logic-0 and the integer -1 to denote a logic-1).

2. Compute a representation of the transformation matrix using the constituent functions.

3. Compute the spectral coefficients between the transformation matrix and the output vector of the function to be synthesized.

4. Choose the largest (in magnitude) spectral coefficient.

5. Realize the function $f_c(x)$ that corresponds to the chosen coefficient in step 4.

6. Compute the error function $e(x) = f_c(x) * f(x)$ with respect to some operator $*$.

7. If $e(x)$ indicates that there are w or fewer errors, go to step 8. Otherwise iterate on the synthesis by going to step 3 and use $e(x)$ as the next function to be synthesized.

8. Combine all the intermediate realizations of the various chosen $f_c(x)$ functions using the $*$ operator and directly realize the function $e(x)$ for the remaining w or fewer errors.

This technique can be used to generate two-level and multi-level tree-type circuits. For two-level realizations, each chosen $f_c(x)$ can be realized in the first stage of the circuit with one multi-input logic gate. The second stage consists of a single combination gate that uses the outputs of all of the chosen constituent functions as its inputs. The circuits resulting from this synthesis technique are *Completely Fanout Free* (CFOF) and have the desirable property of requiring a set of test vectors equal to the number of circuit inputs to test all possible single stuck-at faults [53]. As discussed in [55], the use of spectral design techniques for logic synthesis is known for the ability to produce easily tested circuits. The diagram in Figure 7.5 indicates how the two-level circuit is constructed with each iteration.

If a multi-level circuit is desired the same design procedure is used, but the error function computation is performed slightly differently. The difference is that a new combination gate is used in each iteration. This also allows for changing the operator used to define the error function (*i.e.*, the combination operator) at each iteration. The diagram in Figure 7.6 illustrates the design procedure for a multi-level circuit.

The following theorem states the properties necessary to ensure the convergence of this synthesis algorithm.

Theorem 7.1 *Any given Boolean function, $f(x)$, may be realized with the proposed synthesis technique if the transformation matrix used for the synthesis is of full rank.*

Proof: This proof is a statement that any N-vector can be produced as a combination of a subset of vectors from the set of vectors that are linearly independent over N-space. Each Boolean function to be realized is viewed as an N-vector with components from the binary field. The synthesis procedure described in the preceding text "chooses" a matrix row in each iteration (each row corresponds to a constituent function) to be "combined" with an appropriate combining operator. This process forms the output vector as a combination of row vectors from the transformation matrix. Hence, if the transformation matrix contains at least N rows that span N-space, any function output vector can be realized by a finite number of combined transformation matrix rows. $\square$

Note that this synthesis method can not realize any arbitrary function using a set of constituent functions that do not form a functionally complete set since the resulting transformation matrix will not be of full rank. For example, a

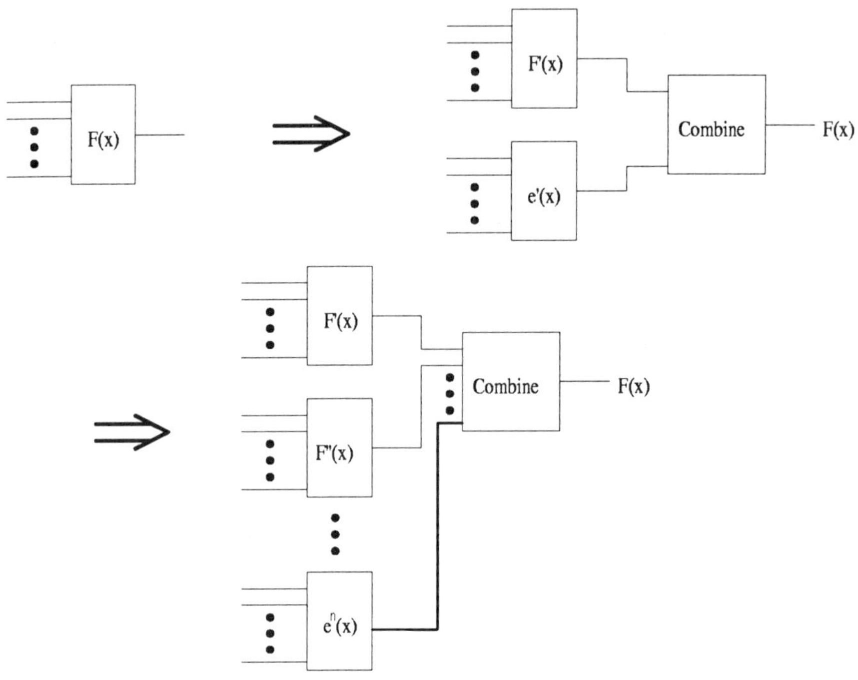

Figure 7.5   Diagram of Two-Level Synthesis Technique

function may not be realized if all constituent functions use only the AND operator.

Synthesis Examples Using the Maximum Correlation Method

In the first example, we will assume that there are no restrictions on the number of inputs per logic gate and on the type of logic gate employed. The first example will show how a two-level circuit can be synthesized. The second example shows how a multi-level circuit can be realized with the constraints that only two-input gates should be used and that the gate type should be OR as much as possible.

Example 7.2 *Consider the realization of the following function*

Table 7.1   Truth Table for the Example Function

x_1	x_2	x_3	F
1	1	1	-1
1	1	-1	1
1	-1	1	-1
1	-1	-1	-1
-1	1	1	1
-1	1	-1	-1
-1	-1	1	-1
-1	-1	-1	1

$$f(x) - \overline{x}_1\overline{x}_3 + x_1\overline{x}_2 x_3 + x_1 x_2 + x_2\overline{x}_3 \tag{7.4}$$

*First, a representation of the function to be synthesized is obtained. Table 7.1
shows the truth table for f.*

*Next, the transformation matrix is computed in terms of OR, AND, and XOR
operations (since there are no constraints on gate types, any matrix could be
used as long as it was of full rank). Also, for each gate type it is possible to
compute $f_c(x)$ functions for all possible combinations of two or more inputs
since we have no restrictions on the maximum number of inputs per gate. Although it is readily apparent that this matrix grows in size with respect to the
size of the x vector, it should be pointed out that the matrix space complexity
is $O(n^2)$ if the design is restricted to two-input gates of type OR, AND, and
XOR. In this example, there are 3 operator types over all combinations of any
two out of n inputs. The total number is*

$$3 \binom{n}{2} = \frac{3}{2}(n)(n-1) = \frac{3}{2}n^2 - \frac{3}{2}n \tag{7.5}$$

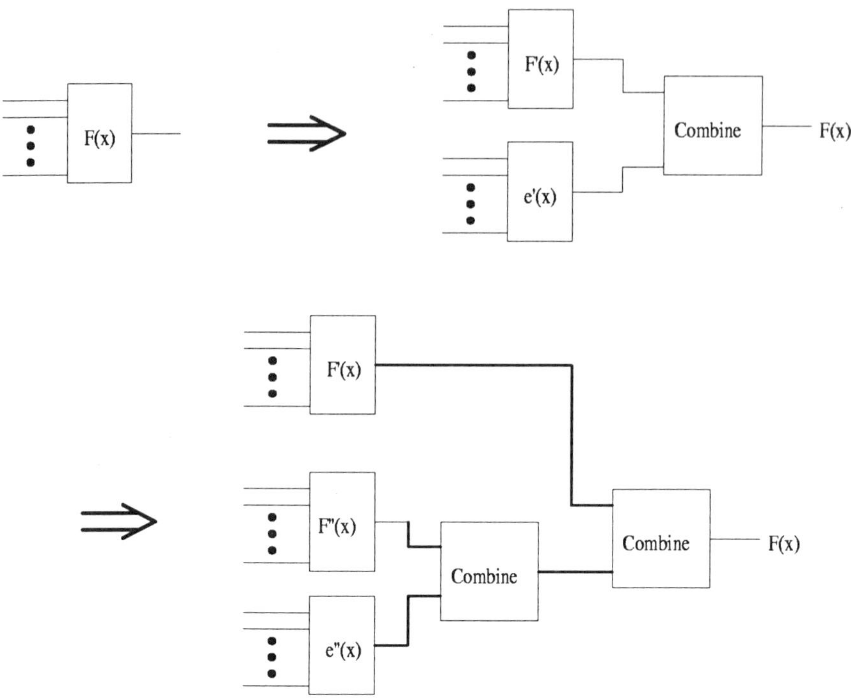

Figure 7.6   Diagram of Multi-Level Synthesis Technique

Added to this value is $n + 1$ additional matrix rows for the computation of the Chow parameters [27, 89] yielding a total number of rows equal to $3/2n^2 - 1/2n + 1 = O(n^2)$ in size complexity of the matrix.

Also, rows are typically included in the transformation matrix that correspond to each component of the x vector and the $f_c(x) = 1$ function (the Chow parameters) in case of high correlation with regard to a single input to $f(x)$. The spectral coefficient $S_f(1)$ indicates the ratio of minterms to the value 2^n of the candidate function $f(x)$. The transformation matrix follows in Figure 7.7 with the symbolic expression corresponding to the constituent function of each row shown to the left.

The resulting spectral coefficient vector with the coefficients in the same order as the constituent functions appear in the matrix is

$$
\begin{array}{r}
1 \\
x_1 \\
x_2 \\
x_3 \\
x_1 \oplus x_2 \\
x_1 \oplus x_3 \\
x_2 \oplus x_3 \\
x_1 \oplus x_2 \oplus x_3 \\
x_1 + x_2 \\
x_1 + x_3 \\
x_2 + x_3 \\
x_1 + x_2 + x_3 \\
x_1 x_2 \\
x_1 x_3 \\
x_2 x_3 \\
x_1 x_2 x_3
\end{array}
\left[
\begin{array}{rrrrrrrr}
1 & 1 & 1 & 1 & 1 & 1 & 1 & 1 \\
1 & 1 & 1 & 1 & -1 & -1 & -1 & -1 \\
1 & 1 & -1 & -1 & 1 & 1 & -1 & -1 \\
1 & -1 & 1 & -1 & 1 & -1 & 1 & -1 \\
1 & 1 & -1 & -1 & -1 & -1 & 1 & 1 \\
1 & -1 & 1 & -1 & -1 & 1 & -1 & 1 \\
1 & -1 & -1 & 1 & 1 & -1 & -1 & 1 \\
1 & -1 & -1 & 1 & -1 & 1 & 1 & -1 \\
1 & 1 & -1 & -1 & -1 & -1 & -1 & -1 \\
1 & -1 & 1 & -1 & -1 & -1 & -1 & -1 \\
1 & -1 & -1 & -1 & 1 & -1 & -1 & -1 \\
1 & -1 & -1 & -1 & -1 & -1 & -1 & -1 \\
1 & 1 & 1 & 1 & 1 & 1 & -1 & -1 \\
1 & 1 & 1 & 1 & 1 & -1 & 1 & -1 \\
1 & 1 & 1 & -1 & 1 & 1 & 1 & -1 \\
1 & 1 & 1 & 1 & 1 & 1 & 1 & -1
\end{array}
\right]
$$

Figure 7.7 Transformation Matrix

$$\overrightarrow{S_f^T}[\{f_c(x)\}] = [-2, -2, 2, -2, 2, -2, 2, -6, 2, -2, 2, 2, -2, -2, -2, -4] \quad (7.6)$$

Since the $f_c(x) = x_1 \oplus x_2 \oplus x_3$ has the largest magnitude spectral coefficient, it is chosen, and the first portion of the circuit is realized with an exclusive-NOR as shown in Figure 7.8.

The error function is computed with respect to an exclusive-OR operator since it is the most robust in terms of the possible operators available for providing the combining stage in the circuit. This robustness property is due to the fact that an XOR can be used to change a 0 to 1 error as well as a 1 to 0 error. The following list describes the properties that determine which gate type may be used as an error operator.

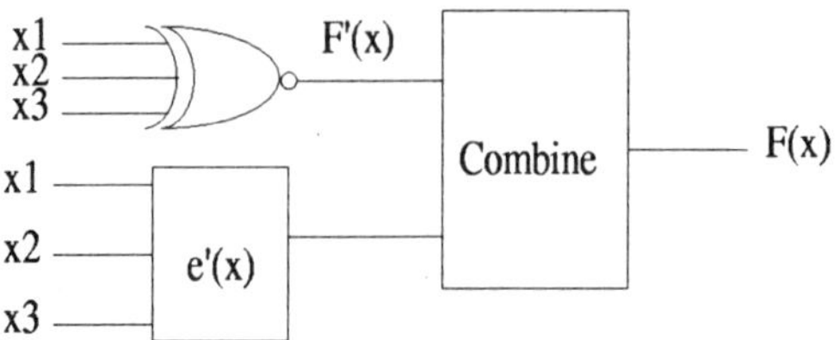

Figure 7.8   First Iteration of Two-Level Synthesis of Example Function

x_1	x_2	x_3	$f(x)$	$f'(x)$	$e(x)$
1	1	1	-1	-1	1
1	1	-1	1	1	1
1	-1	1	-1	1	-1
1	-1	-1	-1	-1	1
-1	1	1	1	-1	1
-1	1	-1	-1	-1	1
-1	-1	1	-1	-1	1
-1	-1	-1	1	1	1

Table 7.2   Comparison of the Error Function to the Example Function

1. *XOR : $x \oplus 1 = \bar{x}$, errors may be 1→0 or 0→1*

2. *AND : $x1 = x$ and $x0 = 0$, all errors must be 1→0*

3. *OR : $x + 1 = 1$ and $x + 0 = x$, all errors must be 0→1*

Returning to the synthesis example, the Table 7.2 is computed. $f'(x)$ refers to the partially synthesized function $f(x)$.

Since there is only a single disagreement between $f(x)$ and $f'(x)$, the terminal condition has been reached, and the remaining term is realized directly. The complete circuit is given in Figure 7.9.

The design of a multi-level circuit with the restriction that all gates are two-input and must be OR type gates as much as possible is given in Example

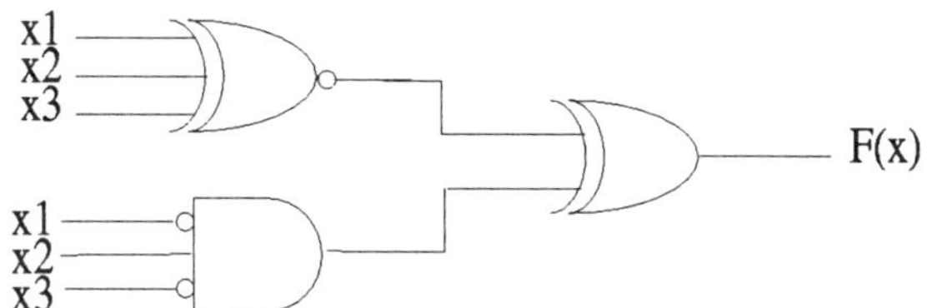

Figure 7.9 Final Circuit using Two-Level Design Process

7.3. The OR operator cannot be used to form a functionally complete set of operators for Boolean algebra. Therefore, at least one other type of gate is needed for the synthesis. This example will also realize the function given in the previous example.

Example 7.3 *The transformation matrix is computed in terms of the Chow parameters and two-input, OR constituent functions only. The following spectrum results:*

$$\overrightarrow{S_f^T}[\{f_c(x)\}] = [-2, -2, 2, -2, 2, -2, 2] \tag{7.7}$$

In this case, all spectral coefficients have equal magnitudes. The constituent function $f_c(x) = x_1 + x_2$ is arbitrarily chosen as a starting point for the synthesis. It is also desirable to use an OR operation for the error function operator since this circuit is to be realized with predominately OR-type gates. By examining the truth table it is seen there are 0 to 1 and 1 to 0 discrepancies which restrict the error operator to be of type XOR. Table 7.3 illustrates the truth table in terms of the function to be realized and the error function.

Next, the spectrum for $e(x)$ is computed, resulting in the following vector of spectral coefficients:

$$\overrightarrow{S_e^T}[\{f_c(x)\}] = [2, 2, -2, -4, -2, -2, -6] \tag{7.8}$$

The constituent function corresponding to -6 is chosen since it is a maximum (in magnitude) which is $f_c(x) = x_2 + x_3$. There is a discrepancy in only one

Table 7.3 Truth Table Contents of the Function and Error Function

x_1	x_2	x_3	$F(x)$	$e(x)$
1	*1*	*1*	*-1*	*-1*
1	*1*	*-1*	*1*	*1*
1	*-1*	*1*	*-1*	*1*
1	*-1*	*-1*	*-1*	*1*
-1	*1*	*1*	*1*	*-1*
-1	*1*	*-1*	*-1*	*1*
-1	*-1*	*1*	*-1*	*1*
-1	*-1*	*-1*	*1*	*-1*

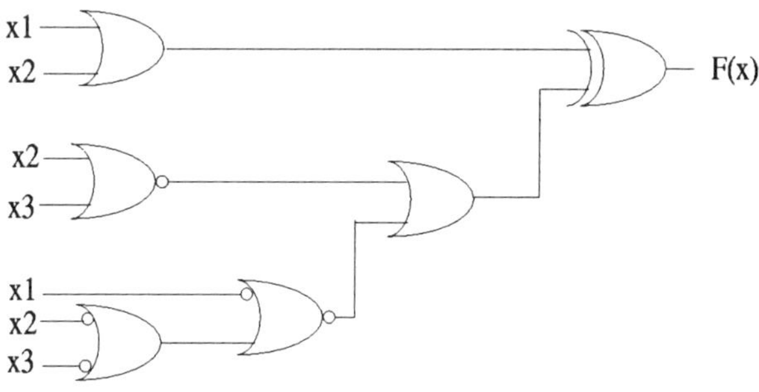

Figure 7.10 Final Circuit using Multi-Level Design Process

place in the truth table, hence we can use a NOR to directly realize this term. The resulting circuit is given in Figure 7.10.

Note that the only real difference between the two-level and multi-level synthesis techniques is in the choice of new error operators at each iteration. This allows for greater flexibility when it is desired to use one gate type as much as possible since that type may be able to be used for the error operator in each iteration. Furthermore, this allows the flexibility to use different error operators at each iteration.

7.2.3 Implementation Issues of the Maximum Correlation Synthesis Method

Repeated vector-matrix multiplications can be very time consuming and inefficient. While this is a convenient way to analyze and understand spectral techniques for Boolean function synthesis, it is generally not the most efficient way to implement these techniques. Using the techniques described in Chapter 5, it is possible to compute spectral coefficients based on cube list or DD based techniques.

If optimization for circuit speed is desired, each spectral coefficient may be weighted by the inverse of the corresponding constituent function delay. This would cause those constituent functions with the least delay and highest correlation to be used.

Existing standard cell logic libraries may also be used with this synthesis technique. All that is required is the output vector of each standard logic cell to be used as a row in the transformation matrix. This allows the synthesis technique presented here to be easily interfaced to existing design environments without the need for changing anything other than the "synthesis engine" itself.

Although this synthesis technique has been developed within the framework of spectral analysis techniques, it may easily be considered to be a method of repeated correlation analysis. The spectral coefficients for a Boolean function have been defined, and various properties pertinent to this method have been derived. This iterative algorithm for the synthesis of combinational circuits based on maximum correlation with constituent functions has been developed and illustrated with examples. The algorithm has been proven to converge when the transformation matrix is of full rank. The proposed method can be custom-tuned to produce circuits with desired properties such as gate count minimization, type of gate, circuit delay minimization, and number of inputs per gate. This result is significant since it allows for complete exploitation of the XOR gate as well as other gate types without resorting to symbolic manipulation of Boolean algebraic equations.

7.3 SYNTHESIS USING SPECTRAL HEURISTICS

A synthesis technique using a subset of spectral coefficients is described in this section. The computation of spectral coefficients using DDs was discussed in Chapter 5 and has been used in the literature [34, 162]. However, much of the attention has been focused on methodologies that require the entire spectrum [57, 94]. Although the computational method presented in [160] reduces the complexity from exponential to polynomial in terms of the number of primary inputs for most functions, the entire spectrum of a function still contains an exponential number of coefficients. This fact provides the motivation for developing a method that uses a subset of spectral coefficients to perform logic synthesis. By using a subset of spectral coefficients, each value may be calculated quickly, and the overall number of computations is no longer exponential.

It has been shown that all 2^n spectral coefficients are required to uniquely represent a Boolean function when the Walsh family of transformation matrices are employed [38, 97]. Thus, a method that uses a subset of coefficients must necessarily employ heuristics since an exact solution cannot be obtained. The use of heuristics in the synthesis of logic functions is very common and has led to some of the most successful tools available today [13, 14, 124].

The method discussed here is developed to produce multilevel circuits that may be optimized for area, device and interconnection minimization. The optimization for timing versus area presents a well known tradeoff. In order to ensure minimal delay, a two-level circuit composed of a maximally reduced set of implicants is the best that can be achieved in terms of critical path length. Alternatively, minimization of area and interconnections generally require a multilevel circuit so that intermediate term sharing between a set of reduced implicants may be exploited.

The approach used here generates a multilevel circuit while also providing shorter paths for more critical inputs (those that become stable later than other inputs). This is reasonable since many real world design problems are specified by considering some valid input signals to be present at the inputs of the circuit before others. With this scheme area optimization is the dominating synthesis parameter; however, some degree of timing constraint can be included by ordering the inputs according to their times of arrival. Finally, circuit testability can be enhanced in this synthesis methodology by producing circuits that minimize internal fanout.

This method is designed to be the intermediate step between a behavioral description of a logic function and an initial circuit structure to use as input to other timing or area optimizers. All circuit minimizers require an initial form of the circuit as input [14, 124], and the final synthesized output can be adversely affected if a highly inefficient input circuit description was provided. Furthermore, the optimizations are usually performed by local changes over portions of the circuit. In contrast, this approach provides an initial circuit structure that was formulated using the global information present in digital logic spectral coefficients.

The remainder of this section is organized as follows. First, the synthesis method is presented, and an examination of the optimization criteria will be provided. Next, the development of the technique is described in detail. The formulation of the spectral heuristics will be explained, and the use of the decomposition methods at each stage of synthesis are discussed. Following the discussion of the philosophy behind the technique, implementation issues will be discussed including the program flow. Finally, some examples of this method are presented.

7.3.1 Description of the Heuristic Synthesis Methodology

This synthesis technique produces a circuit by determining an output gate first and working back toward the inputs. The output gate is chosen by using the information contained in the subset of spectral coefficients commonly referred to as the Chow parameters [27]. Based upon the properties of the Chow parameters, a set of heuristic rules are applied to choose the appropriate gate. The heuristics have been formulated such that the chosen gate will be maximally correlated to the entire function, and hence, the remaining portion of the function will be simplified. In order to take advantage of the efficient method for computing single spectral coefficients described in Chapter 5, the intermediate functions as well as the initial input function are represented in terms of BDDs.

At each stage of the synthesis, once the output gate is chosen, at least one primary input is removed from each remaining intermediate function. Therefore, the range of the intermediate function always decreases by at least one half. The particular input that is chosen to be removed is determined by the optimization criteria. If timing optimizations are desired, the slower arriving inputs are removed first resulting in fewer gates in their propagation path. If area and interconnection minimization are required, the input is chosen using the principle of maximal subfunction independence resulting in the intermediate functions being as simple as possible. Since at least one primary input is discarded at each step in the processing flow of the synthesis technique, convergence is guaranteed.

The reason the Chow parameters were chosen rather than some other higher ordered spectral coefficient is that a correlation measure relative to a specific primary input is needed. A higher ordered coefficient would give a correlation measure relative to a specific function of a subset of the primary inputs. In fact, there are cases when higher ordered coefficients are useful as secondary measures such as those used in [159] to detect the dependence of a structural circuit on various parity functions.

7.3.2 Optimization Criteria

In the past, area minimization has generally been measured as the number of implicants in a minimized cover of a function, and the minimization of gates and interconnections has been measured as the total number of literals in a minimized cover [13]. The most common way of incorporating this type of optimization has been by using the concepts of "don't cares" [14, 136]. Another popular method that has been exploited by many researchers is the use of "permissible functions" [124].

The minimization criteria used in this implementation is the creation of intermediate functions that are as degenerate as possible. Since a single primary input is guaranteed to be removed at each stage of the synthesis, the resulting intermediate functions will contain at least one less primary input. However, if the intermediate functions also become degenerate, additional inputs may be discarded resulting in significantly simpler functions remaining to be realized. Furthermore, the determination of which, if any, of the inputs are redundant is implicitly achieved since the intermediate functions are represented by BDDs formed by applying the *RESTRICT* operation on the original BDD [17]. The *RESTRICT* operation returns a BDD with a logic 1 or 0 substituted for all instances of a specified input variable. This occurs because a BDD is defined as a BDD with a specific variable ordering that has been maximally reduced [17]. Thus, a redundant input cannot be contained in a BDD. The test for maximum subfunction degeneracy validates the use of the heuristics for choosing the primary input to decompose about since it is guaranteed to produce subfunctions that depend on fewer primary inputs rather than a random selection of a primary input.

In addition to the exploitation of intermediate function degeneracy, interconnection optimization is achieved through the structure that a circuit synthesized by this technique must have. Since each stage allows a single primary input to be discarded from the next synthesis step, a characteristic overall circuit structure results. Primary inputs are discarded through the use of the Shannon decomposition [147] in most cases. Various forms of this decomposition formula imply the structure of each intermediate portion of the resulting circuit. Figure 7.11 depicts three possible forms for a single iteration of this synthesis technique. By choosing forms that incorporate a fanout of two, or, by even restricting those forms to no fanout, interconnections can be minimized and they are also local to the current area of the circuit that is being synthesized. This method prevents gates near the input side of the circuit from directly driving gates near the output end of the circuit. Thus, gates that are not close together are decoupled within the resulting circuit minimizing interconnection complexities.

Although the prototype implementation of this method concentrated on the incorporation of area minimization, a timing optimization criteria could be incorporated by exploiting the order in which primary inputs are discarded from each intermediate function. When the designer supplies the BDD of the circuit to be synthesized, timing information must also be specified. Specifically, the inputs must be grouped into classes that are ranked by the speed at which they will appear at the inputs to the resulting circuit. Each class may contain a single input implying a strict timing order of arrival or, at the other extreme, a single class may be specified inferring that all signals will be present at the same time. The specification of these classes in effect dictate the delay versus area tradeoffs in the final result. If a strict ordering is supplied, the synthesizer is forced to discard the primary inputs from the intermediate functions in a specific order. Therefore, area minimization is achieved only through the choosing of the particular output gate at each stage. It should be noted that the heuristics that are used to determine the output gates were developed with area minimization in mind so that a strict ordering does not necessarily cause the resulting circuit to be overly large; it just removes a degree of freedom by not allowing the synthesizer to choose the most prudent input to discard. Alternatively, if all inputs are in the same class, implying that all will arrive at the input of the circuit simultaneously, the synthesizer is allowed to choose the input to discard that will most likely result in an intermediate function with a high degree of redundancy as well as to choose which type of output gate to use. The practical way to specify the timing classes is to order only those inputs that are critical in terms of arrival time and to place all others in the same class. This will not only allow the synthesizer to enforce the timing criteria but also give it maximum freedom for minimizing the resultant area.

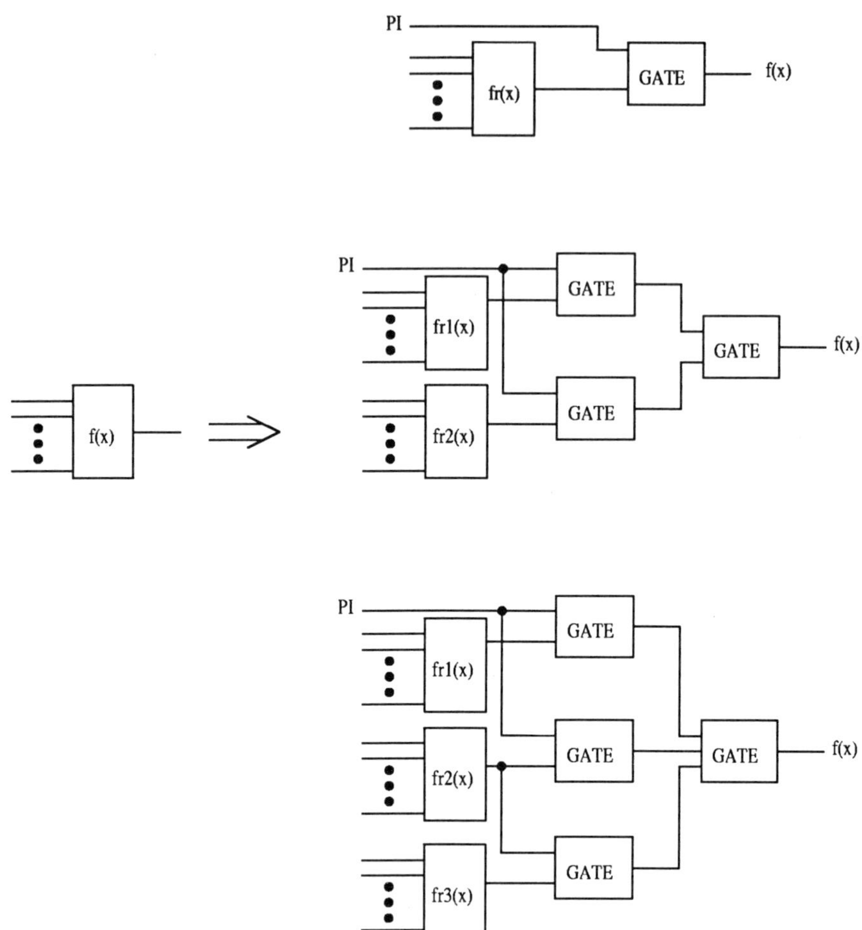

Figure 7.11   Diagram of a Single Iteration of the Heuristic Synthesis Method

The third optimization criteria is that of testability. As is the case with the iterative synthesis technique described in the previous section, this method produces CFOF circuits that are highly testable since they only require a number of test vectors to detect any single stuck-at fault in the circuit equal to the number required to test for a single stuck-at fault at the primary inputs [53]. By choosing the decompositions at each stage of the synthesis to be such that no fanout is generated, the resulting circuit will have no internal fanout and be highly testable.

This synthesis methodology employs a set of heuristics based upon properties of the Chow parameters. The set of heuristics is used to choose the output gate at each level of synthesis. The basis for the heuristics for the choice of the output gate was the examination of the Chow parameters for all possible Boolean functions of two variables. It is essential that the correct gate be chosen when only two primary inputs remain in order to ensure that the synthesis algorithm terminates and does not oscillate when this terminal condition occurs. The algorithm is thus guaranteed to converge since a primary input is discarded at each intermediate stage. At the the last iteration, when only two primary inputs remain, a gate is guaranteed to be chosen that will result in termination of the algorithm. The following section will describe the details concerning the development of these heuristics, and it will list them in a table.

The other main idea in the implementation is the maximal redundancy test used to determine which primary input to discard. Since the function to be realized is in BDD form, it is very efficient to apply the BDD *RESTRICT* operation for an input variable. When the *RESTRICT* operation is applied the returned BDD will always depend on one less variable, and in many cases several other inputs will also become redundant. The maximal redundancy algorithm computes the BDDs for the restriction of each input and chooses the input that results in the most redundancy.

7.3.3 Development of the Heuristic Technique

The input to the synthesis program is a BDD representing the circuit to be synthesized. A queue is maintained that points to each intermediate BDD to be synthesized. Initially, the BDD of the entire function is placed in the queue. At each stage of the synthesis a BDD is popped from the queue. If the BDD depends on 2 or more inputs, the Chow parameters are computed. Based upon the Chow parameter heuristics, an output gate is chosen. Once the output gate is chosen, the primary input to discard from the remainder functions must be obtained. If there is a timing optimization, the primary input corresponding to the largest arrival time is chosen. Otherwise, the maximal redundancy test is applied to choose the appropriate primary input to be discarded.

Formulation of the Heuristics

The heuristics were derived by observing the Chow parameters for all 16 possible Boolean functions of 2-variables and by exploiting the properties of spectral coefficients. The set of rules are organized in a hierarchical manner so that the rules providing the simplest residual BDDs are chosen first. Table 7.4 contains the list of heuristics and rules used in the synthesis tool. The value σ is defined as the sum of the first order spectral coefficients as given in Equation 7.9 and f_i^0, f_i^1 represent the Shannon cofactors, $f_0 = f(x_1, \ldots, x_{i-1}, 0, x_{i+1}, \ldots, x_n)$ and $f_1 = f(x_1, \ldots, x_{i-1}, 1, x_{i+1}, \ldots, x_n)$.

$$\sigma = \sum_{i=1}^{n} S(x_i) \tag{7.9}$$

To illustrate how the heuristics were chosen consider the Boolean functions and their associated Chow parameters given in Table 7.5. There are 16 entries in Table 7.5 corresponding to all possible functions of 2 variables.

As an example, consider the function $x_1 + x_2$. The zero order spectral coefficient $S(0)$ is less than 0, and the sum of the first order coefficients σ is greater than 0. Therefore, whenever $S(0) < 0$ and $\sigma > 0$, the OR gate is chosen as the dominant function.

Shannon Decomposition Forms

Once the dominant output gate is chosen, the function must be decomposed into two residual functions. The decomposition method used in this synthesis tool is based on variations of the Shannon decomposition formula when the chosen output gate is not the XOR or XNOR. The Shannon form was chosen since at least one primary input is guaranteed to be discarded. In order to accommodate the various output gate forms, the Shannon decomposition formula was rearranged into several forms. A complete table of the decomposition types used is given in Table 7.6. It should also be noted that a recent paper

Table 7.4 Heuristics and Rules for the Synthesis Methodology

main heuristic	secondary heursitic	function choice
$\|S(0)\| = 2^n$	$S(0) < 0$	$f = 1$
	$S(0) > 0$	$f = 0$
$\|S(x_i)\| = 2^n$	$S(x_i) < 0$	$f = \overline{x}_i$
	$S(x_i) > 0$	$f = x_i$
$\|S(x_i)\| = 2^n - \|S(0)\|$	$S(0) < 0$ and $S(x_i) < 0$	$f = \overline{x}_i + f_i^1$
	$S(0) < 0$ and $S(x_i) > 0$	$f = x_i + f_i^0$
	$S(0) > 0$ and $S(x_i) < 0$	$f = \overline{x}_i \cdot f_i^0$
	$S(0) > 0$ and $S(x_i) > 0$	$f = x_i \cdot f_i^1$
$S(0) > 0$	$\sigma < 0$	NOR
	$\sigma > 0$	AND
	$\sigma = 0$ and $S(x_1) \geq 0$	AND
	$\sigma = 0$ and $S(x_1) < 0$	NOR
$S(0) < 0$	$\sigma < 0$	NAND
	$\sigma > 0$	OR
	$\sigma = 0$ and $S(x_1) \geq 0$	OR
	$\sigma = 0$ and $S(x_1) < 0$	NAND
$S(0) = 0$	$S(x_i) = 0 \forall i$	XOR
	$\sigma > 0$	AND
	$\sigma < 0$	OR
	$\sigma = 0$ and $S(x_1) \geq 0$	XNOR
	$\sigma = 0$ and $S(x_1) < 0$	XOR

Table 7.5 Chow Parameters for all Boolean Functions of 2 Variables

function	chow parameters			σ	gate
	$S(0)$	$S(x_1)$	$S(x_2)$		
0	4	0	0	4	constant 0
$x_1 x_2$	2	2	2	6	AND
$x_1 \overline{x}_2$	2	2	-2	2	NOR
x_1	0	4	0	4	literal x_1
$\overline{x}_1 x_2$	2	-2	2	2	NOR
x_2	0	0	4	4	literal x_2
$x_1 \oplus x_2$	0	0	0	0	XOR
$x_1 + x_2$	-2	2	2	2	OR
$\overline{x_1 + x_2}$	2	-2	-2	-2	NOR
$\overline{x_1 \oplus x_2}$	0	0	0	0	XNOR
$\overline{x}_2$	0	0	-4	-4	literal $\overline{x}_2$
$x_1 + \overline{x}_2$	-2	2	-2	-2	OR
$\overline{x}_1$	0	-4	0	-4	literal $\overline{x}_1$
$\overline{x}_1 + x_2$	-2	-2	2	-2	OR
$\overline{x_1 x_2}$	-2	-2	-2	-6	NAND
1	-4	0	0	-4	constant 1

has illustrated other relationships between Walsh spectral coefficient types and Boolean function decompositons [62].

When the output gate is chosen as the XOR or XNOR, a spectral based decomposition is attempted first. Usually the spectral based decomposition is very effective in terms of partitioning the primary inputs; however, in those cases where inferior decompositions are achieved, rules based on various Shannon decompositions are used. When the Shannon forms are used, the form that contains residual BDDs with the smallest number of dependent variables is chosen.

The idea behind the spectral based decomposition method is to determine two partitioned subfunctions such that one depends only on highly correlated primary inputs while the other depends upon inputs with a small correlation. When the dominant gate is chosen to be the XOR, the function f is partitioned in the form as shown in Equation 7.10. The subfunction g is formed by evaluating f with all highly correlated inputs set to a logic 0 as given in Equation 7.11.

$$f = g \oplus h \tag{7.10}$$

$$g = f(0, 0, 0, \ldots, x_{n-i}, x_{n-i+1}, \ldots, x_n) \tag{7.11}$$

The choice of the highly correlated inputs is made by choosing those inputs whose corresponding spectral coefficients have a magnitude greater than $|2^{n-1}|$. Mathematically stated, the criteria for choosing the highly correlated inputs is given in expression 7.12. Once g is computed, the corresponding h subfunction is obtained directly by evaluating Equation 7.13.

$$|S_f[x_i]| > 2^{n-1} \rightarrow \; use \; 0 \; for \; x_i \tag{7.12}$$

$$h = f \oplus g \tag{7.13}$$

Table 7.6   Shannon Decomposition Forms

type	form	residuals	number of gates
OR	$\overline{x}_i f_0 + x_i f_1$	f_0, f_1	3
	$f_0 f_1 + \overline{x}_i f_0\overline{f}_1 + x_i\overline{f}_0 f_1$	$f_0 f_1, f_0\overline{f}_1, \overline{f}_0 f_1$	4
	$\overline{\overline{f}_0 + \overline{f}_1} + \overline{x}_i\overline{(\overline{f}_0 + f_1 + x_i f_0 + \overline{f}_1)}$	$\overline{f}_0 + \overline{f}_1, \overline{f}_0 + f_1, f_0 + \overline{f}_1$	4
	$f_0 f_1 + [f_0\overline{f}_1 \oplus x_i(f_0 \oplus f_1)]$	$f_0 f_1, f_0\overline{f}_1, f_0 \oplus f_1$	3
	$\overline{\overline{f}_0 + \overline{f}_1} + [(\overline{\overline{f}_0 + f_1}) \oplus x_i(f_0 \oplus f_1)]$	$\overline{f}_0 + \overline{f}_1, \overline{f}_0 + f_1, f_0 \oplus f_1$	3
	$f_0 f_1 + [\overline{f}_0 f_1 \oplus \overline{x}_i(f_0 \oplus f_1)]$	$f_0 f_1, \overline{f}_0 f_1, f_0 \oplus f_1$	3
	$(\overline{\overline{f}_0 + \overline{f}_1}) + [(\overline{f}_0 + \overline{f}_1) \oplus \overline{x}_i(f_0 \oplus f_1)]$	$\overline{f}_0 + \overline{f}_1, f_0 + \overline{f}_1, f_0 \oplus f_1$	3
AND	$(x_i + f_0)(\overline{x}_i + f_1)$	f_0, f_1	3
	$(\overline{\overline{f}_0\overline{f}_1})(x_i + \overline{\overline{f}_0 f_1})(\overline{x}_i + \overline{f_0\overline{f}_1})$	$\overline{f}_0\overline{f}_1, \overline{f}_0 f_1, f_0\overline{f}_1$	4
	$(f_0 + f_1)(x_i + f_0 + \overline{f}_1)(\overline{x}_i + \overline{f}_0 + f_1)$	$f_0 + f_1, f_0 + \overline{f}_1, \overline{f}_0 + f_1$	4
	$\overline{\overline{f}_0\overline{f}_1}[(x_i \oplus \overline{\overline{f}_0 f_1}) + (\overline{f_0 \oplus f_1})]$	$\overline{f}_0\overline{f}_1, \overline{f}_0 f_1, \overline{f_0 \oplus f_1}$	3
	$(f_0 + f_1)\{[x_i \oplus (f_0 + \overline{f}_1] + (\overline{f_0 \oplus f_1})\}$	$f_0 + f_1, f_0 + \overline{f}_1, \overline{f_0 \oplus f_1}$	3
	$(\overline{\overline{f}_0\overline{f}_1})[(\overline{x}_i \oplus \overline{f_0\overline{f}_1}) + (\overline{f_0 \oplus f_1})]$	$\overline{f}_0\overline{f}_1, f_0\overline{f}_1, \overline{f_0 \oplus f_1}$	3
	$(f_0 + f_1)\{[\overline{x}_i \oplus (\overline{f}_0 + f_1)] + (\overline{f_0 \oplus f_1})\}$	$f_0 + f_1, \overline{f}_0 + f_1, \overline{f_0 \oplus f_1}$	3
NOR	$\overline{\overline{x}_i\overline{f}_0 + x_i\overline{f}_1}$	$\overline{f}_0, \overline{f}_1$	3
	$\overline{\overline{f}_0\overline{f}_1 + \overline{x}_i\overline{f}_0 f_1 + x_i f_0\overline{f}_1}$	$\overline{f}_0\overline{f}_1, \overline{f}_0 f_1, f_o\overline{f}_1$	4
	$\overline{f_0 + f_1 + \overline{x}_i(\overline{f_0 + \overline{f}_1}) + x_i(\overline{\overline{f}_0 + f_1})}$	$f_0 + f_1, f_0 + \overline{f}_1, \overline{f}_0 + f_1$	4
	$\overline{\overline{f}_0\overline{f}_1 + [f_0 f_1 \oplus x_i(f_0 \oplus f_1)]}$	$\overline{f}_0\overline{f}_1, \overline{f}_0 f_1, f_0 \oplus f_1$	3
	$\overline{(\overline{f_0 + f_1}) + [(f_0 + \overline{f}_1) \oplus x_1(f_0 \oplus f_1)]}$	$f_0 + f_1, f_0 + \overline{f}_1, f_0 \oplus f_1$	3
	$\overline{\overline{f}_0\overline{f}_1 + [f_0\overline{f}_1 \oplus \overline{x}_i(f_0 \oplus f_1)]}$	$\overline{f}_0\overline{f}_1, f_0\overline{f}_1, f_0 \oplus f_1$	3
	$\overline{(\overline{f_0 + f_1}) + [(\overline{f}_0 + f_1) \oplus \overline{x}_i(f_0 \oplus f_1)]}$	$f_0 + f_1, \overline{f}_0 + f_1, f_0 \oplus f_1$	3
NAND	$\overline{(x_i + \overline{f}_0)(\overline{x}_i + \overline{f}_1)}$	$\overline{f}_0, \overline{f}_1$	3
	$\overline{(\overline{f_0 f_1})(\overline{x}_i + \overline{f}_0 f_1)(x_i + f_0\overline{f}_1)}$	$f_0 f_1, \overline{f}_0 f_1, f_0\overline{f}_1$	4
	$\overline{(\overline{f}_0 + \overline{f}_1)(\overline{x}_i + f_0 + \overline{f}_1)(x_i + \overline{f}_0 + f_1)}$	$\overline{f}_0 + \overline{f}_1, f_0 + \overline{f}_1, \overline{f}_0 + f_1$	4
	$\overline{(\overline{f_0 f_1})[f_0\overline{f}_1 \oplus x_i(f_0 \oplus f_1)]}$	$f_0 f_1, f_0\overline{f}_1, f_0 \oplus f_1$	3
	$\overline{(\overline{f}_0 + \overline{f}_1)[(\overline{f}_0 + f_1) \oplus x_i(f_0 \oplus f_1)]}$	$\overline{f}_0 + \overline{f}_1, \overline{f}_0 + f_1, f_0 \oplus f_1$	3
	$\overline{(\overline{f_0 f_1})[\overline{f}_0 f_1 \oplus \overline{x}_i(f_0 \oplus f_1)]}$	$f_0 f_1, \overline{f}_0 f_1, f_0 \oplus f_1$	3
	$\overline{(\overline{f}_0 + \overline{f}_1)[(f_0 + \overline{f}_1) \oplus \overline{x}_i(f_0 \oplus f_1)]}$	$\overline{f}_0 + \overline{f}_1, f_0 + \overline{f}_1, f_0 \oplus f_1$	3
XOR	$\overline{x}_i f_0 \oplus x_i f_1$	f_0, f_1	3
	$f_0 \oplus x_i(f_0 \oplus f_1)$	$f_0, f_0 \oplus f_1$	2
	$f_1 \oplus \overline{x}_i(f_0 \oplus f_1)$	$f_1, f_0 \oplus f_1$	2
XNOR	$\overline{\overline{x}_i\overline{f}_0 \oplus x_i\overline{f}_1}$	$\overline{f}_0, \overline{f}_1$	3
	$\overline{\overline{f}_0 \oplus x_i(f_0 \oplus f_1)}$	$\overline{f}_0, f_0 \oplus f_1$	2
	$\overline{\overline{f}_1 \oplus \overline{x}_i(f_0 \oplus f_1)}$	$\overline{f}_1, f_0 \oplus f_1$	2

7.3.4 Heuristic Synthesis Method Implementation

This synthesis technique was implemented using the *C* programming language. The basic data structure used in the implementation is a first-in first-out (FIFO) queue that contains pointers to the intermediate BDDs. Initially the queue is initialized to point to the BDD representing the function to be realized. At each stage of the synthesis, the BDD pointed to by the pointer at the top of the FIFO is operated upon and an output gate is chosen for implementation. If any remainder BDDs are created, pointers corresponding to them are inserted at the tail of the FIFO. The synthesis is complete when the FIFO becomes empty indicating the entire circuit has been built.

Examples and Results

Several of the *ISCAS85* and *IWLS* benchmark circuits were synthesized using this technique. In order to determine the relative effectiveness of this technique, the *IWLS* benchmarks were synthesized using the *mis*II [14] tool from Berkeley and mapped to a small cell library identical to one used for the heuristic method.

Two different *mis*II scripts were used to synthesize these benchmarks. The first script consisted of an *espresso* minimization [13] followed by the *mis*II *simplify* and *sweep* commands. The second *mis*II script omitted the *espresso* simplification and consisted of the *collapse*, *simplify*, and *sweep* commands. Table 7.7 contains a summary of these results. The column labeled '*mis*II/*espresso* Size' contains the number of logic gates *mis*II required when the first script was used as input. The fifth column labeled '*mis*II Size' contains the number of gates the resulting circuit required when the *espresso* minimization was not used. Finally, the column labeled 'Heuristic Size' contains the number of logic gates in the resulting circuit using the synthesis method just described.

This comparison was chosen because the prototype synthesis tool described here is currently implemented to operate on single output functions only. By selecting outputs from the benchmarks in Table 7.7 that result in non-trivial single-output functions and synthesizing with both our method and *mis*II, an evaluation of the effectiveness of the approach is given.

These results are mixed in that the method does significantly better and significantly worse in several cases. For those circuits whose Shannon cofactors about some input variable become very small (depend on as few as possible

Table 7.7 Comparison of Spectral Based Heuristic Method with *mis*II

Circuit Name	Output Name	Number Inputs	*mis*II Size	*mis*II/*espresso* Size	Heuristic Method Size
5xp1	*output* 1	7	14	12	25
9sym	*output* 1	9	262	103	194
alu4	*output* 2	14	21	26	13
bw	*output* 1	5	65	41	14
con1	*output* 1	6	8	6	11
duke2	*output* 4	22	263	122	11
misex3	*output* 1	14	96	373	16
rd53	*output* 1	5	18	7	9
rd73	*output* 1	7	114	116	125
rd84	*output* 1	8	98	213	253
sao2	*output* 4	10	43	73	11
vg2	*output* 2	25	35	82	46
apex1	*output* 12	45	96	74	38
misex3c	*output* 1	14	34	28	34
xor5	*output* 1	5	4	54	4
clip	*output* 4	9	81	97	170

Table 7.8 Experimental Results of the Spectral Based Heuristic Logic Synthesizer

Circuit	Output	Inputs	Size
c432	329gat	27	11009
c880	879gat	44	174
c880	863gat	36	113
c880	768gat	10	17
c2670	401	11	20
c3540	399	26	60
c3540	364	25	29
c5315	1002	9	15
c5315	871	24	81
c5315	661	24	49

inputs), the method does quite well. This is because the principle of maximum redundancy is incorporated into our method. Also, circuits with XOR operations tend to do well in this method since techniques for detecting the XOR relationship [159] are incorporated into the tool.

Table 7.8 contains a summary of the results obtained when significantly larger circuits are synthesized. The first column contains the name of the benchmark circuit. The remaining columns contain the name of the output, the number of primary circuit inputs, and the resulting circuit size in terms of logic gates using the same cell library as in the previous results.

In cases where the synthesis technique performed poorly, the results turn out to be trees of Shannon decompositions, although they are present in a mixture of forms (depending on the one chosen from Table 7.6), with subsequent variable dependencies of size $n-1$ where the previous stage depends on n variables. This is the worst case scenario since all functions can be realized as trees of Shannon decompositions, but such realizations require large gate counts. This result is not overly surprising since this method was designed to exploit redundancy through the principle of maximum subfunction independence as described in a previous section. Due to the good performance on some circuits and the very poor performance on others, this method can be used as an optimization step in a larger synthesis tool made up of other optimization routines, or it can be used to initially transform a circuit in BDD form into a netlist suitable for input to some other synthesis tool.

In terms of timing and memory requirements, all of the above data was generated using a 100 MHz Sun SPARCstation 20 with 160 MB of internal RAM. Although the total memory usage for each example was not explicitly measured, none of the synthesis runs above resulted in page swapping, and the wall clock times averaged from a few seconds to several minutes. For those examples with extremely large gate counts, the run times were much larger. As an example, the extreme case of c432, output 329gat required 10 hours to complete.

This method allows for the transformation of a function in BDD form to a structural netlist without using an intermediate flattening process. Furthermore, for some classes of circuits this transformation has very good area minimization properties. The area optimization results are mixed. Some benchmark circuits compare favorably with a standard synthesis tool while others require a significantly larger gate count. The circuits that yield small gate counts with this technique generally have a high degree of maximum subfunction independence as described in this book.

7.4 TECHNOLOGY MAPPING

Technology mapping [40, 57, 100] is a critical step in logic synthesis that can significantly affect the quality of the final circuit. A common approach to technology mapping is to partition the circuit and to then identify a library cell for each subcircuit. The matching of a candidate cell to a library cell is a major computational step in technology mapping.

7.4.1 Boolean Matching

Recent interest has centered upon technology mapping methods based on *Boolean matching* [21, 25, 32, 96, 112, 141, 142, 143, 172], an approach which depends solely on the functionality of the subcircuit and the library cell and not on their structures. The key to such an approach is to determine that two functions are equivalent relative to a permitted set of operations. Here we consider two functions to be matchable if they are equivalent with respect to input negation, input permutation and output negation: the so-called *NPN equivalence*.

Table 7.9 Function Classification Statistics

n	2	3	4	5	6
function	16	256	65,536	$\approx 4.3 \times 10^9$	$\approx = 1.8 \times 10^{19}$
NPN	4	14	222	616,126	$\approx 2 \times 10^{14}$
spectral	2	3	8	48	-

The NPN equivalence of Boolean functions was first considered in 1965 by Harrison [84]. In particular, the numbers of NPN equivalence classes for $n \leq 6$ are known (see Table 7.9).

The approach presented here [118] selects an appropriate permutation and negations to map a given function to a novel canonical form. Two functions are mapped to the same canonical form if, and only if, they are in the same NPN class. The combination of the permutations and negations required to map the functions to the canonical form can be combined to map one function to the other.

The approach here is distinct from those Boolean matching techniques that directly map one function to another. With such methods, a given subcircuit must be tested against each of a set of library cell functions. While filters can be applied to reduce the number of full comparisons undertaken, this approach can be computationally very expensive. With the method considered here, the canonic form can be stored with each library cell, and after mapping the function from the subcircuit to the canonical form only a simple test for equivalence is required. Hashing techniques can be used for greatest efficiency.

Spectral techniques have been applied to function classification [92] (see Table 7.9) where the spectral translations given in Table 3.1 are employed. While a much more compact function classification scheme is achieved, it is not of interest for technology mapping since the exclusive-OR operations required for input and output translation are prohibitive. Here we use spectral techniques but address only NPN equivalence.

The approach here is similar to those in [32] and [170], but we use more properties of the spectrum and a new canonical form rather than direct function matching. The approach is also related to the one in [143] in that it makes heavy use of function symmetry.

7.4.2 The Canonical Form

We consider a function $f(x_1, x_2, \ldots, x_n)$ with Walsh spectrum $\mathbf{S}$ under S-encoding. Since each Boolean function has a unique Walsh spectrum, the notion of functional equivalence classes extends directly to spectral equivalence classes. The canonical form identifies one particular spectrum (function) in each class.

Let

$$v_p = \sum_{\forall \alpha \subseteq \beta} s_{p\alpha}^2, \quad 1 \le p \le n, \quad \beta = (1, \ldots, p-1, p+1, \ldots, n) \tag{7.14}$$

v_p is the sum of the squares of those spectral coefficients with p as a subscript.

It is easily shown [92] that $f(x_1, x_2, \ldots, x_n)$ is symmetric in x_i and x_j if, and only if, $s_{i\alpha} = s_{j\alpha}$ for all $\alpha \subseteq (1, \ldots, i-1, i+1, \ldots, j-1, j+1, \ldots, n)$. We employ the term *negative symmetry* for the situation where $f(x_1, x_2, \ldots, x_n)$ is invariant under the interchange of x_i and $\overline{x}_j$, or equivalently the interchange of $\overline{x}_i$ and x_j. In this case $s_{i\alpha} = -s_{j\alpha}$. Note that $v_i = v_j$ if $f(x_1, x_2, \ldots, x_n)$ is symmetric or negative symmetric in x_i and x_j, but the converse does not always hold. A set of inputs which are in pairs symmetric or negative symmetric will be termed a *symmetry set*.

NPN Canonical Form: The NPN canonical spectrum in the NPN spectral equivalence class $\mathbf{E}$ is the spectrum $\mathbf{S}$ which satisfies

1. $s_0 \ge 0$.

2. $s_i \ge 0, \quad 1 \le i \le n$.

3. The input variables satisfy the following ordering rules:

    (a) $v_i \ge v_{i+1}$, $1 \le i \le n-1$ so that $\{x_1, x_2, \ldots, x_n\}$ may be partitioned into $P_1, P_2, \ldots$ such that all variables in the same P_i have the same v value.

    (b) Each P_i is then partitioned into an appropriate number of symmetry sets $Q_1^i, Q_2^i, \ldots$ which are ordered by decreasing cardinality *i.e.* $|Q_j^i| \ge |Q_{j+1}^i|$ for all j.

    (c) For every $|Q_j^i| = |Q_{j+1}^i|$, $|s_\alpha| \ge |s_{\hat{\alpha}}|$, for the first α such that $|s_\alpha| \ne |s_{\hat{\alpha}}|$ where $\hat{\alpha}$ is the rearrangement of α found by the pairwise interchange of the variables of Q_j^i and Q_{j+1}^i.

4. Given $\mathbf{C} \subseteq \mathbf{E}$ the set (including $\mathbf{S}$) of all spectra satisfying criteria (1) to (3), $\mathbf{S}$ is such that for the first β (in Hadamard coefficient order) where $s_\beta < 0$, there is no $\hat{\mathbf{S}} \in \mathbf{C}$ such that $\hat{s}_\gamma \geq 0$, for all $\hat{s}_\gamma$ up to and including $\hat{s}_\beta$ (in Hadamard order).

5. Given $\mathbf{D} \subseteq \mathbf{E}$ the set (including $\mathbf{S}$) of all spectra satisfying criterion (4), $\mathbf{S}$ is such that for all other $\hat{\mathbf{S}} \in \mathbf{D}$, $s_\delta > 0$ for the first $\hat{s}_\delta$ after $\hat{s}_\beta$ (in Hadamard order) such that $s_\delta \neq \hat{s}_\delta$.

Criterion (3) separates the symmetry sets into two categories: those whose positions are fixed and those which are interchangeable with a neighboring set. We term the latter *interchangeable* sets. Note that while two interchangeable sets have corresponding coefficients of equal magnitude, they can not have strictly equal coefficients or strictly negated coefficients since that would indicate a symmetry or inverse symmetry and the two sets would have been combined. Also note that interchangeable sets may come in larger groupings than pairs but they must appear in sequence in the same P_i. Criteria (4) and (5) identify one spectrum from all possible arrangements of the interchangeable symmetry sets.

Theorem 7.2 *The spectrum satisfying criteria (1) to (5) is unique in* $\mathbf{E}$.

Proof: Let $\mathbf{C}$ be the set of all spectra in $\mathbf{E}$ satisfying criteria (1) to (3). Clearly $\mathbf{C}$ is nonempty since a spectrum in $\mathbf{E}$ either satisfies criteria (1) and (2) or can be mapped to a spectrum in $\mathbf{E}$ which is accomplished by suitable function and input negations. Similarly, a spectrum satisfying criteria (1) and (2) either satisfies criterion (3) or it can be mapped to one in $\mathbf{E}$ that does by input permutation.

Suppose $|\mathbf{C}| > 1$ and choose any two spectra $\mathbf{S}^1, \mathbf{S}^2 \in \mathbf{C}$. Since $\mathbf{S}^1$ and $\mathbf{S}^2$ are in the same NPN equivalence class, they have equivalent symmetry sets up to variable relabeling. Criterion (3) prescribes a unique ordering of these symmetry sets with the exception of the interchangeable symmetry sets discussed above. Hence, the ordering prescribed by (3) is unique up to absolute value of the coefficients and it follows that $|s_\alpha^1| = |s_\alpha^2|$ for all α.

If $\mathbf{S}^1$ and $\mathbf{S}^2$ both satisfy criterion (4), they clearly must do so at the same coefficient position β. However, they can not then both satisfy criterion (5) since there must be a δ later in Hadamard than β such that s_δ^1 and s_δ^2 have different signs, otherwise they would be the same spectrum. It follows that there is exactly one spectrum in $\mathbf{E}$ which satisfies all five criteria. $\square$

It is straightforward to apply the given criteria and determine whether a given spectrum is in NPN canonical form. Transforming a given spectrum to canonical is more difficult.

7.4.3   The Boolean Matching Method

The interaction between the permutation and negation operations complicates the determination of the canonical form. For example, $x + \overline{y}$ and $\overline{x} + y$ are NPN equivalent, but one has the choice of permuting the variables or negating them. Inverting the exclusive-OR function, or one of its inputs, yields the same result, and inverting the inputs and the output of a self-dual function yields the original function. There are many much more complex situations. As a result it is not possible to consider permutation and negation separately.

There is a total of $2(n!)2^n$ NPN transformations that can be applied to an n variable function. For $n = 4, 8$, and 10 this yields 768, 20,643,840 and 7,431,782,400, respectively. Clearly the search space must be pruned as much as possible. Our method has two phases:

Phase I: If $s_0 < 0$, we apply function negation. After that we apply input negation for all i where $s_i < 0$. With some care these translations can be applied in one pass through the coefficients. During that same pass we compute the v_i.

Once the v_i are known, we examine input variable pairs for symmetry or inverse symmetry. Note that x_i and x_j need only be considered if $v_i = v_j$.

Next the input variables are permuted to satisfy criterion (3) of the canonical form. Criterion (3a) and (3b) simply require appropriate sorting. Criterion (3c) is more problematic since reordering a pair of variables can affect the coefficient magnitude relationships for other variable pairs. We use a simple interchange sort which is repeated until no changes are required. It is easily verified that this process will terminate. $\hat{\mathbf{S}}$ will denote the spectrum resulting from phase I.

Phase II: We next search a set of candidate spectra to determine the canonical form. Each candidate is $\hat{\mathbf{S}}$, with particular set of negations applied, including the empty set, *i.e.* $\hat{\mathbf{S}}$ is a candidate itself and is the first one considered. The negations available are the function, if $\hat{s}_0 = 0$, and each input such that $\hat{s}_i = 0$. In theory, we must try all negation patterns, 2^{n+1} in the worst case. Limiting the search space is considered below.

It is easily shown [92] that all s_i and s_0 are nonzero if the function truth vector has an odd number of ones. In this case, no negation search is required.

After applying the appropriate negations to form the candidate, we must order its interchangeable symmetry sets (if it has any). This is done so that within P_i, each pair of adjacent interchangeable sets are such that for x_k, the first variable in Q_j^i, and x_t, the first variable in Q_{j+1}^i, $s_{k\alpha}$ is positive and $s_{t\alpha}$ is negative (they by definition have the same magnitude) for the first α such that $s_{k\alpha} \neq s_{t\alpha}$. Such an α must exist or the function would be symmetric in the two variables. Once again the reordering of a pair can affect other pairs so the interchange process must be iterated until no changes are required.

During the search we keep track of the *best* candidate considered so far which is the one that is better than all the others thus considered according to criteria (4) and (5) of the normal form. $\hat{S}$ with its interchangeable symmetry sets properly ordered is the initial best candidate. At the end of the search, the best candidate found is the canonical form.

As we form each candidate by applying negations to $\hat{S}$, a process which we do in Hadamard order, we perform a check at the first negative valued coefficient in the candidate. If that coefficient is at a position earlier than the first negative valued coefficient of the best candidate to date, and it does not involve any variables in interchangeable symmetry sets, the candidate being formed can be rejected without completing its construction or reordering its interchangeable sets. The experimental results shown in the next section show the rejection process is quite effective.

7.4.4 Experimental Results

Boolean Matching

The method described above has been implemented as a C program. Table 7.10 illustrates how the method performs for n from 4 to 10. For $n = 3D4$, all 65,536 functions were used. For $n > 4$, 100,000 randomly generated functions were used.

Table 7.10 Spectral matching program statistics

n	functions searched	%	full searches	rejected searches	reject. %	max.	avg.	2^{n+1}	msec. avg.
4	29,851	45.55	83,944	77,486	47.99	9	4.5	32	0.10
5	43,113	43.11	129,264	50,768	28.20	10	4.2	64	0.19
6	39,657	39.66	115,153	30,238	20.80	8	3.7	128	0.36
7	35,223	35.23	95,565	18,038	15.88	7	3.2	256	0.72
8	30,886	30.89	78,684	10,623	11.89	7	2.9	512	1.53
9	25,831	25.83	62,001	6,051	8.89	7	2.6	1,024	3.27
10	21,473	21.47	48,933	3,850	7.29	8	2.5	2,048	7.30

Table 7.10 reports the number of functions requiring negation while searching and the total percentage considered for each n. The remaining functions were found to be canonical after phase I of the method as either the zero and first order coefficients were all positive or the coefficients were all nonnegative. Note that in the former case, the interchangeable subsets must be ordered, but that is not necessary when all coefficients are nonnegative.

Full searches is the total number of candidates completely considered for all functions and *rejected searches* is the number of candidates rejected by the first negative coefficient method described above. The rejection percentage is with regard to the total number of searches, full plus rejected. The table gives the maximum and average number of candidates considered per function. The maximum possible number of candidates, 2^{n+1}, is shown for comparison. The final column shows the average CPU time per function (user time as reported by the UNIX time function) for tests run on a Sun SPARCServer 20.

Several observations can be made. The percentage of functions requiring negation searching drops dramatically with n. We must be cautious on this point since as n increases, we are sampling a much smaller fraction of the function space (there are 2^{2^n} functions of n variables). However, the data lends support to the supposition that most functions are easy to put into canonical form and only a relatively small number require extensive searching. This is certainly supported by our experience with the exhaustive analysis for $n = 4$.

The total candidates number drops accordingly, but somewhat surprisingly, so does the rejection percentage. It is not clear why this should happen. The maximum number of candidates considered for a function stays fairly consistent, whereas the average number drops somewhat. Both are well below the worst case number of candidates and show the overall effectiveness of the method.

Finally, we note that the time required to put a function into canonical form is not high even though our program has not been highly optimized.

Technology Mapping

The spectral approach to Boolean matching has been used in a technology mapping procedure targeted to filed programmable gate arrays developed by X. Wang [168]. Wang's approach uses the well-known SIS system [145] to perform technology independent optimization to produce a fine-grain representation of the Boolean network where each node in the optimized network has, at most, two product terms and, at most, two literals. The fine granularity is required for best results in the partitioning phase of the mapper.

The first step of the mapper is to partition the network into a set of single output subnetworks. In fact, the partitioning results in a forest of trees. Redundancies are then eliminated. Next, each of the trees is decomposed such that each node in the tree has, at most, two children, and the Boolean function associated with each node is either a two-input AND or a two-input OR, or NOT in the case of a node with only one child. Partitioning is done to this level since all cell libraries will include these basic functions, and therefore a solution will be guaranteed.

After partitioning and decomposition, the forest of trees is covered by an interconnection of cells from the technology cell library while optimizing some cost function. Wang used simple logic block count as the cost function, since in an FPGA logic blocks are the resource to be assigned to the realization of Boolean functions.

Wang's method uses standard dynamic programming and cluster functions [56] used in many approaches to technology mapping. A cluster function is a Boolean function that corresponds to a set of connected nodes. The idea is to cover all nodes in the forest by cluster functions; each of which can be realized by a cell in the target technology library.

There is not uniqueness in Wang's method of verifying that a possible target function matches a cell in the technology library. This is where the spectral approach to Boolean matching is used.

The technology cell library is preprocessed to identify the available NPN equivalence classes and the spectral canonical form for each. During the technology mapping covering process, a target cluster function is put into spectral canoni-

Table 7.11   Statistics for the ACT-1 Technology Library

Inputs	1	2	3	4	5	6	7	8	Total
Realizable Functions	2	8	47	210	285	128	21	1	702
Number of NPN Classes	1	2	8	43	75	37	9	1	176
Functions in largest NPN Class	2	6	11	19	16	8	4	1	-

cal form, and hashing techniques are used to verify if there is a match between the cluster function and a cell in the library. This process is made even faster by using v_p as given in equation 7.14 to identify which cells are possible matches, and indeed if there is a possible match, before putting the cluster function into canonical form.

As an indication of the appropriateness of this approach, Table 7.11 shows statistics for the Actel Act-1 technology library. That library is based on a physical cell that can realize 702 functions distributed over n as shown. Using Boolean matching based on NPN classes greatly reduces the searching required. For example for $n = 4$, 43 NPN canonic functions need be considered rather than 210. If only permutation of inputs is to be considered, *i.e.* inverters are to be minimized, at most 19 functions must be considered after the NPN match is made. That consideration can lead to a significant speed-up of the technique by using spectral negation properties.

The important thing to note regarding the Boolean matching in this approach to technology mapping is that it is not just the cluster functions appearing in the final solution that need to be matched. It is all target cluster functions encountered in the dynamic programming search for a best solution. Hence, the performance of the mapper is greatly enhanced by an approach that speeds up the matching process. The results presented by Wang [168], show that, for standard benchmark problems, his mapper produces technology mappings with comparable cell counts to other mappers reported in the literature, but often much faster. Also, Wang's results show the time required for the technology mapping phase for his approach are at least an order of magnitude, and sometimes two, less than the time required for the technology independent optimization.

The quality of Wang's results is a result of the partitioning, decomposition and covering techniques used. The efficiency is a result of using the spectral approach to Boolean matching. Wang's experiments were for the relatively simple ACT-1 library. The approach will show even better performance as the size and complexity of the library increases.

7.5 SUMMARY

Four applications of spectral methods for logic synthesis have been presented. Spectral translation properties are those that classify functions into partitions characterized by all functions having the same spectral coefficient values, but in a different order. The spectral translation property of "input translation" has been applied in different ways by several researchers resulting in a partitioning of a function into a linear n-input, n-output function and a remainder function.

Two heuristic methods were presented that synthesize logic functions based on the values of sets of spectral coefficients. In the first case, the principle of maximum correlation was used to decompose a target function into an interconnected set of constituent functions represented by transformation matrix rows. The second method utilized the zero and first order spectral coefficients with a set of heuristics and decomposition formulae to realize a netlist.

Finally, spectral based techniques for technology mapping were described. Technology mapping is a critical process in standard-cell library based synthesis techniques.

8

LOGIC VERIFICATION

Logic verification is to check the correct behavior of a given circuit against its specification. This specification can be given in the form of another circuit at some higher level of abstraction or as a description of properties that the circuit is to obey. In this chapter a review of some "classical" approaches for verification is given followed by a discussion of a spectral based technique for equivalence checking [154]. For more detailed surveys of verification methodologies see [31, 79, 88, 99].

8.1 CLASSICAL APPROACHES

In the following the equivalence checking problem for two Boolean functions of n variables $f(X)$ and $g(Y)$ is discussed. This problem can be viewed as being comprised of two sub-problems: establishing the correspondence of input and output variables for the two circuit descriptions and, once the correspondence is established, determining if the two circuit abstractions have the same functional behavior for all possible input values.

The first problem of input and output signal correspondence is a special case of the Boolean matching problem and thus has applications in both logic synthesis and verification. Chapter 7 has described spectral methods for Boolean matching in general. The use of "signatures" is one way to address the input/output correspondence problem. A signature is some structure associated with each input and output of a candidate circuit. Many times signatures are numeric values that can be associated with spectral coefficients. If the signatures are unique and easily computed, the correspondence problem is easily solved.

The "traditional" way to make sure that two designs represent the same function is to use simulation. Dependent on the background information known for the circuit, specific stimuli can be selected or random input patterns may be used otherwise. However, due to the increasing complexity of modern designs, this approach is yielding less reliable results since it is impractical to examine a significant portion of the overall design space of the two designs. Full equivalence checking of two n-input designs (with known input/output signal correspondence) requires 2^{n+1} simulations and 2^n comparisons of model responses. In many cases exhaustive simulation is infeasible.

Several techniques based on symbolic methods, such as the use of BDDs, have been proposed, and also the combination of BDDs and simulation has been suggested [72]. Another alternative is to consider word-level functions, which are functions with a Boolean range and an integer domain. Based on these definitions, extensions of BDDs can be defined that are well suited for representing arithmetic functions like multipliers [19, 45]. These approaches are encouraging since they allow close integration of the verification of bit-level circuits and high-level descriptions such as those specified using a *Hardware Description Language* (HDL) that often make use of word-level operations. However, for large circuits with several million components, there is little hope for finding a complete representation using a single type of DD.

8.2 SIGNATURE METHODS

In this section the relationship between signature-based approaches and spectral coefficient matching are examined. Signature methods are used to correlate the inputs and outputs of specifications of a circuit. This situation arises in the technology mapping phase for automated synthesis of standard cell designs and can also occur during equivalence checking when the correspondence between circuit inputs and outputs are lost due to tool relabeling or some other interaction by the designer.

In general, these two situations have differing constraints. The Boolean matching problem encountered in logic synthesis must determine an input and output variable correspondence among a relatively large library of cells and a candidate circuit; however, the sizes of the two circuits are relatively small and, polarity changes in the inputs or outputs are typically allowed. For verification (equivalence checking), two specific abstractions of a circuit are given so there is no need to search among a large library; however, the sizes of the circuits

can generally be quite large. Also, the input correspondences among the two
abstractions must be identified. It is possible for the labels in the two circuit descriptions to be different for the identical circuit input. In this case,
some other means of identifying input correspondence must be used. If the two
abstractions do indeed differ in functionality, depending on the particular signature used, it may be impossible to find the correspondences since the circuits
are truly different in a functional sense.

In the remainder of this section the input/output correspondence problem is examined from a verification point of view since Chapter 7 discussed the Boolean
matching problem using spectral coefficients. Here, we do not offer a new approach to the correspondence problem, rather the intent is to show that previous
techniques are in fact spectral approaches.

The use of "satisfy-count" signatures has been discussed in [26, 107, 120, 121].
These types of signatures are computed as the number of all possible variable
assignments that cause a single output function (or some sub-function) to evaluate to a logic-1 value. When the candidate functions are expressed in DD
form, these signature functions are easily computed. In [26, 107, 120], it was
proposed that satisfy-counts of positive cofactors be used as signatures.

Using the notation of Chapter 2, it is noted that the satisfy-count of a positive
cofactor about variable x_i of a single output, n-input function f, can be ex-
pressed as $2^n \wp\{f|x_i\}$ where the conditional probability $\wp\{f|x_i\}$ is the fraction
of times that the cofactor f_{x_i} evaluates to a logic-1. Assuming that f is fully
specified (*i.e.* $\wp\{x_i\} = \frac{1}{2}$) and using Bayes' rule as given in Equation 2.8, it is
seen that the satisfy-count signature of a cofactor about variable x_i (which is
denoted by SC_i) can be expressed in terms of output probabilities as given in
Equation 8.1.

$$SC_i = 2^{n+1} \wp\{f \cdot x_i\} \tag{8.1}$$

Equations 5.7 and 5.8 relate the output probability quantities $\wp\{f\}$ and $\wp\{f \cdot x_i\}$ with the zero and higher ordered Walsh spectral coefficients, respectively.
Combining Equations 8.1 and 5.8 it is seen that the relationship between satisfy-
counts of positive cofactors and Walsh spectral coefficients is given as:

$$SC_i = 2^n - \frac{1}{2}S_f(x_i) + 2^n \wp\{f\} \tag{8.2}$$

The Walsh spectral coefficient, $S_f(x_i)$, is a first-ordered spectral coefficient using S-encoding. The quantity $\wp\{f\}$ is related to the satisfy-count SC_0 of the complete function f as $SC_0 = 2^n \wp\{f\}$. Using Equation 5.7 it is easily seen that $SC_0 = 2^{n-1} - \frac{1}{2}S_f(0)$. Substituting these results into Equation 8.2 yields

$$SC_i = SC_0 + 2^n - \frac{1}{2}S_f(x_i) \qquad (8.3)$$

$$SC_0 = 2^{n-1} - \frac{1}{2}S_f(0) \qquad (8.4)$$

These results indicate that the use of satisfy-count signatures for the input correspondence problem or for Boolean matching are equivalent to the use of Walsh spectral coefficients such as was reported in [34].

8.3 SUBSETS OF HAAR COEFFICIENTS FOR EQUIVALENCE CHECKING

In this section a probabilistic equivalence checking method is developed based on the use of partial *Haar Spectral Diagrams* (HSDs). Partial HSDs are defined and used to represent a subset of Haar spectral coefficients for two Boolean functions. The resulting coefficients are then used to compute and to iteratively refine the probability that two functions are equivalent. The method described here can be useful for the case where two candidate functions require extreme amounts of memory for a complete BDD representation. Experimental results are provided to validate the effectiveness of this approach.

8.3.1 Motivation for the Use of Haar Coefficients in Equivalence Checking

The technique described here allows for the equivalence checking problem to be formulated in terms of a subset of Haar spectral coefficients [91, 92]. Given a set of Haar spectral coefficients, we examine the probability that $f(X) = g(Y)$. This allows the equivalence checking problem to be iteratively refined in terms of possible error by accounting for the existence of more matching coefficients. Thus, techniques that provide subsets of Haar spectral coefficients [65, 83, 152] for representations of f and g can be used for non-tautology checking. A similar approach using an arithmetic transform and a decision diagram structure known as an *Interleaved BDD* (IBDD) has also been proposed [94]. The technique described here differs due to the fact that we utilize partial HSDs versus IBDDs allowing us to make use of the multi-resolution, normalized Haar wavelet transform [91, 92] rather than the arithmetic transform. This allows one to obtain the Haar spectral coefficients directly from a traversal of the HSD without performing additional spectral computations. Furthermore, the multi-resolution nature of the Haar transform offers advantages in the probability calculations since higher ordered coefficients can represent disjoint portions of the function of interest.

In this approach we adapt the method reported in [83] that allows the Haar spectral coefficients to be represented as an HSD with the concept of the partial BDD as given in [134]. This allows for a partial function representation to be used for quickly computing subsets of Haar spectral coefficients avoiding problems that may arise for functions that result in very large BDDs when represented in their fully specified form. Once the subsets of Haar spectral coefficients are found to be equivalent for two candidate functions, f and g, we compute the probability that f and g are equivalent. If any two same-ordered Haar spectral coefficients are found that have different values, we can declare that $f \neq g$ and halt the process.

A discussion of the background of partial BDDs and HSDs is reviewed followed by a section on the mathematical basis of the technique. The mathematical basis includes a review of relevant aspects of the Haar transform and contains the derivations for the probability computations. Next, we present a simple example and the results of some preliminary experiments that indicate the effectiveness of using matching Haar coefficients for statistical verification. Finally, a section containing conclusions is given.

Incomplete Construction

As long as symbolic simulation can be carried out completely, the verification process succeeds. But problems arise if BDDs do not fit in the main memory of a computer. This might be due to several reasons.

The first (and simplest) reason is that a "bad" variable ordering has been chosen. Furthermore, the ordering in which the operands are combined is very important, as can be seen by the following simple example:

Example 8.1 Let $\mathcal{F}$ be an AND gate with three inputs f, g and h that occurs during symbolic simulation of a circuit. BDD packages based on recursive synthesis have to compute:

$$(f \cdot g) \cdot h, \; f \cdot (g \cdot h) \text{ or } (f \cdot h) \cdot g$$

The order in which the calculation is performed largely influences the number of nodes that are needed during the computation, *e.g.* if $f \cdot g$ is computed first, but $h = 0$. In this case the result $f \cdot g$ (which might be large) is computed first even though the result of the AND gate is 0.

Some first steps for finding good orderings involve traversing gates in circuit representations as described in [50]. However, there also exist functions for which the corresponding BDD size becomes exponential (independent of the variable ordering). The best known example is the multiplier [16].

In [134] an approach based on *partial information* has been proposed. If the BDD size becomes too large some parts can be projected to a new "terminal" node, called U for undefined. The drawback of this method is that the complete functionality of the represented circuit is no longer present and complete verification is not possible. Fortunately, the resulting partial DD gives enough information to compute at least some Haar coefficients. This fact will allow us to formulate the equivalence checking technique based on matching subsets of Haar coefficients using only the partial BDDs.

To give a better understanding of partial BDDs, including the value U, we consider the function from [134] given by the table in Figure 8.1. As can be seen, the BDD for this function requires 6 non-terminal nodes (in the following, we fix the variable ordering). We assume that the memory of the BDD package is limited to 4 non-terminal nodes. Thus, complete construction is not possible. However, if two runs are made using partial information and the rest is projected to an undefined value, U (see Figure 8.2 and 8.3), we see that the complete function can be obtained using two partial BDDs.

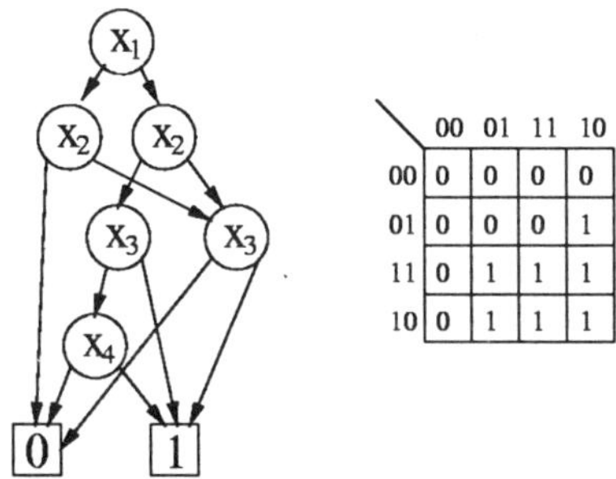

Figure 8.1 Complete BDD

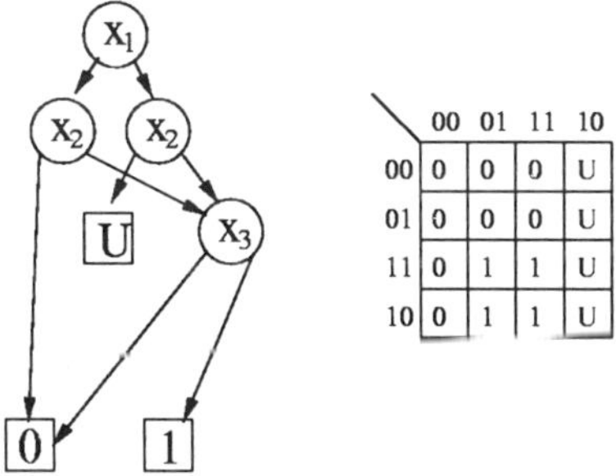

Figure 8.2 First incomplete BDD

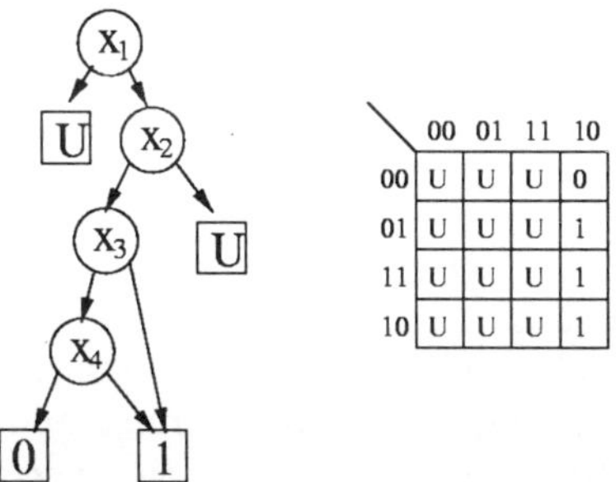

Figure 8.3 Second incomplete BDD

Haar Spectral Diagrams

In [83] a directed graph referred to as a *Haar Spectral Diagram* (HSD) is defined
that represents the Haar spectrum of a Boolean function. Although parts of

this material have been discussed in an earlier chapter, an overview of the important points as related to this technique is included here for completeness.

HSDs are isomorphic to BDDs (with the exception that all BDD terminal vertices are "mapped" to a common HSD terminal vertex). This allows the BDD representation of a function to double as a representation of the Haar spectrum with extra memory storage required only in the form of an additional edge-attribute value. The additional storage is needed because all 1-edges in the HSD have a Haar spectral coefficient as an attribute.

The enabling observation for defining the HSD is that the Haar transformation matrix can be expressed in terms of Kronecker products if the natural order of the coefficients is permuted. The n-dimensional transformation matrix that produces the coefficients in the permuted order T^n can be represented as a sum of matrices denoted as A^n and D^n as given in Equation 8.5.

$$T^n = A^n + D^n \tag{8.5}$$

A^n can now be defined by the Kronecker product relation (denoted by the $\otimes$ operator) as:

$$A^n = \begin{bmatrix} 1 & 0 \\ 0 & 1 \end{bmatrix} \otimes A^{n-1} + \begin{bmatrix} 0 & 0 \\ 1 & -1 \end{bmatrix} \otimes D^{n-1} \tag{8.6}$$

The initial cases are $A^0 = 0$ and $D^0 = 1$. It is observed that the first row of A^n is all zeros and only the first row of D^n is non-zero, thus the spectral vector due to T^n can be represented by the two vectors due to A^n and D^n separately. By using this observation and viewing a non-terminal node of a BDD as pointing to two disjoint subfunctions, we can represent the spectrum of the subfunctions (the subfunction spectra are actually scaled by a constant in this case) as two different portions of the entire vector due to T^n. Figure 8.4 is similar to the diagram originally appearing in [83] and illustrates this relationship.

Using these observations, it is possible to represent the Haar spectrum of a function by annotating all 1-edges of the graph (and the pointer to the initial node) with Haar spectral coefficients.

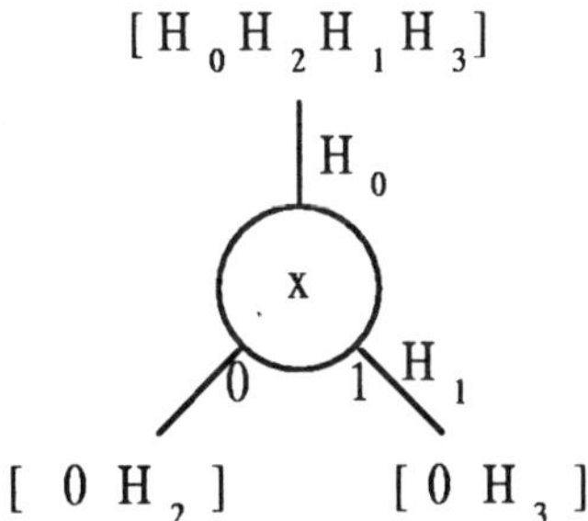

Figure 8.4   Non-terminal in HSD

As an example, consider the Boolean function, $f = x_1 x_2 + \overline{x}_3$. Equation 8.7 gives the Haar spectrum for this function. If the T^3 matrix were used, the resulting permuted spectral vector would become:

$$[H_0, H_1, H_2, H_5, H_1, H_6, H_3, H_7] = [-2, -2, 0, -2, 2, -2, 2, 0]$$

These coefficients can be used as annotation values on the Shannon decision tree representing the example function as shown in Figure 8.5. As is well known, the corresponding BDD can be formed by removing all isomorphic subgraphs and redundant nodes from the decision tree representation. If these reductions are carried and the edge annotations are retained, the BDD/HSD results as shown in Figure 8.6.

$$
\begin{bmatrix}
1 & 1 & 1 & 1 & 1 & 1 & 1 & 1 \\
1 & 1 & 1 & 1 & -1 & -1 & -1 & -1 \\
1 & 1 & -1 & -1 & 0 & 0 & 0 & 0 \\
0 & 0 & 0 & 0 & 1 & 1 & -1 & -1 \\
1 & -1 & 0 & 0 & 0 & 0 & 0 & 0 \\
0 & 0 & 1 & -1 & 0 & 0 & 0 & 0 \\
0 & 0 & 0 & 0 & 1 & -1 & 0 & 0 \\
0 & 0 & 0 & 0 & 0 & 0 & 1 & -1
\end{bmatrix}
\begin{bmatrix}
-1 \\ 1 \\ -1 \\ 1 \\ -1 \\ 1 \\ -1 \\ -1
\end{bmatrix}
=
\begin{bmatrix}
-2 \\ 2 \\ 0 \\ 2 \\ -2 \\ -2 \\ -2 \\ 0
\end{bmatrix}
\qquad (8.7)
$$

The entire spectrum can be recovered through traversals of the HSD, and several properties arise due to the BDD reduction rules. As an example, any 1-edge that is annotated by a negative value must point to either another non-terminal node or the logic-0 terminal. Likewise, a positive valued 1-edge attribute im-

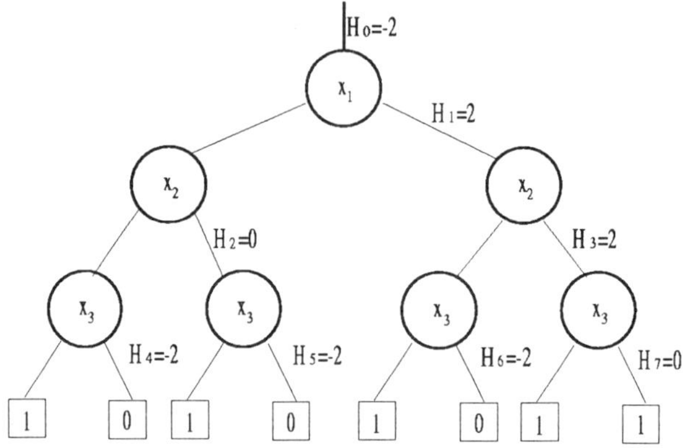

Figure 8.5 Shannon Decision Tree with Haar Coefficient 1-edge Annotations

plies the reverse of this rule. Furthermore, if a 1-edge attribute is 0-valued, it cannot point to a terminal node since it would be removed by the deletion rule. With knowledge of these properties and a given HSD, the entire spectrum can easily be reconstructed.

Using the example HSD in Figure 8.6, it is immediately apparent that $H_0 = -2$, $H_1 = 2$ and $H_3 = 2$ since these values are explicit 1-edge attributes. It is also inferred that $H_2 = 0$ and $H_4 = H_5 = -2$ since all traversals from the x_1 0-edge do not encounter any x_2 node. $H_6 = -2$ by the attribute value on the 1-edge from the x_3 node also since this edge is traversed for the path where $x_1 = 1$, $x_2 = 0$ and $x_3 = 1$. Also $H_7 = 0$, since no 1-edge from x_2 points to x_3, the rightmost x_3 node was removed from the Shannon tree due to its redundancy.

8.3.2 Mathematical Basis and Derivation

In this section the notation used throughout the remainder of the chapter is defined, and relations between probabilistic events and Haar spectral coefficients are derived.

Notation

The following notation is used:

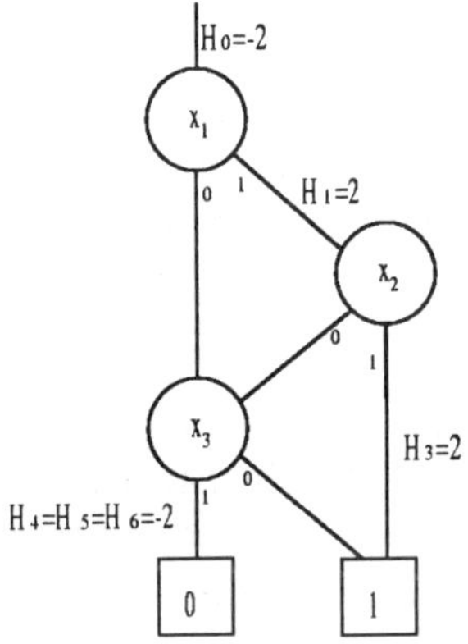

Figure 8.6   Example HSD

- $\overrightarrow{H}^T$ represents the transpose of the normalized Haar spectral coefficient vector representing some function $f(X)$.

- $H_i(f)$ represents the individual i^{th} Haar spectral coefficient of the Boolean function $f(X)$ where $\overrightarrow{H}^T = (H_0, H_1, \ldots, H_{2^n-1})$. H_i is also represented as H_s^o in some of the literature where o is the order of the spectral coefficient and s is the sth Haar function [92].

- $\wp[A]$ is the discrete probability that some event A occurs.

- $\wp[f]$ is the output probability of a Boolean function f which is the likelihood that $f = 1$ given the distribution of the dependent variables in X.

- S_i is the event that $H_i(f) = H_i(g)$, that is, the ith Haar spectral coefficients of f and g are equal in value.

- E is the event that $f(X) = g(Y)$, that is, the functions f and g are functionally equivalent.

Haar Spectrum

Although, techniques for the computation of the Haar spectrum have been described previously in Chapter 5 and [152], an overview of the more important aspects of this type of computation is included here for completeness.

Each transformation matrix row consists of the integer elements -1, +1 and 0. An integer -1 represents the Boolean 1 constant, an integer +1 represents the Boolean 0 constant, and an integer 0 indicates the absence of a Boolean constant. Each row represents a particular normalized Haar function f_c dependent upon n or fewer variables where n is the number of variables of f, the function to be transformed.

Figure 8.7 contains the normalized Haar transformation matrix for any function of $n = 3$ variables. It is noted that higher ordered coefficients are computed from matrix row functions with a decreasing range space dimension. In fact, this decrease in the dimension of the range space corresponds directly to various Shannon cofactors of the function to be transformed.

The output vector of the function to be transformed generally contains integers with -1 representing logic-1 and +1 representing logic-0. With this viewpoint we can define the number of matches between a particular transformation matrix row vector as the number of times the row vector and function vector components are simultaneously equal to -1 or +1. Since some of the rows represent functions that are masked by cofactors, the row-function space is less than 2^3 in size and the presence of a 0 value acts as a place holder.

$$
\begin{array}{r}
f \\
x_1 \\
x_2 \cdot f_{\overline{x}_1} \\
x_2 \cdot f_{x_1} \\
x_3 \cdot f_{\overline{x}_1 \overline{x}_2} \\
x_3 \cdot f_{\overline{x}_1 x_2} \\
x_3 \cdot f_{x_1 \overline{x}_2} \\
x_3 \cdot f_{x_1 x_2}
\end{array}
\begin{bmatrix}
1 & 1 & 1 & 1 & 1 & 1 & 1 & 1 \\
1 & 1 & 1 & 1 & -1 & -1 & -1 & -1 \\
1 & 1 & -1 & -1 & 0 & 0 & 0 & 0 \\
0 & 0 & 0 & 0 & 1 & 1 & -1 & -1 \\
1 & -1 & 0 & 0 & 0 & 0 & 0 & 0 \\
0 & 0 & 1 & -1 & 0 & 0 & 0 & 0 \\
0 & 0 & 0 & 0 & 1 & -1 & 0 & 0 \\
0 & 0 & 0 & 0 & 0 & 0 & 1 & -1
\end{bmatrix}
$$

Figure 8.7   Normalized Haar Transformation Matrix for $n = 3$

The presence of cofactors in the Haar constituent functions can be accounted for by using Bayes' theorem to represent these quantities as output probabilities of the AND of the function to be transformed with its respective dependent literals. These functions are shown to the left of the transformation matrix in Figure 8.7.

In order to determine the total number of matching outputs between f and a row-function, it is necessary to determine when both simultaneously evaluate to a logic-0 level as well as a logic-1 level. We denote the percentage of the total number of matches of logic-0 between some f and a row-function as p_{m0} and likewise for the logic-1 levels, p_{m1}. With this viewpoint the composite f_c expressions can be constructed (shown to the left of the transformation matrix in Figure 8.7) that utilize cofactors of the function to be transformed to restrict the range space and to dictate where the relative location of the valid output of the f_c function occurs in the 2^n row vector components.

Given these observations, we see that the kth normalized Haar spectral coefficient can be calculated as

$$H_k = 2^{n-i}[2(p_{m0} + p_{m1}) - 1] \tag{8.8}$$

Where n is the dimension of the range space of the function to be transformed, f, and i is the dimension of the range space of a particular Shannon cofactor of f. If N_m represents the number of times an intermediate product value of $+1$ occurs in the computation of a particular normalized Haar spectral coefficient (corresponding to 1×1 and -1×-1 products) and N_{mm} corresponds to the number of times a product value of -1 occurs (corresponding to 1×-1 products), then the kth normalized Haar spectral coefficient is given as

$$H_k = N_m - N_{mm} \tag{8.9}$$

It is noted that the sum of N_m and N_{mm} must necessarily equal 2^{n-i} where i indicates the number of variables about which cofactors have been taken. Substituting this observation into Equation 8.9 yields

$$H_k = 2N_m - 2^{n-i} \tag{8.10}$$

We define p_m to be the total fraction of times that a matching output between the f and f_c functions occur, therefore $p_m = N_m/2^{n-i}$. Furthermore, $p_m = p_{m0} + p_{m1}$. Substituting these definitions into Equation 8.10 yields the result

Table 8.1   Relationship of the Haar Spectrum and Output Probabilities

Symbol	i	$n-i$	p_{m1}	p_{m0}
H_0	0	3	$\wp[f \cdot 0]$	$\wp[\overline{f} \cdot \overline{0}]$
H_1	0	3	$\wp[f \cdot x_1]$	$\wp[\overline{f} \cdot \overline{x}_1]$
H_2	1	2	$\dfrac{\wp[f \cdot \overline{x}_1 \cdot x_2]}{\wp[\overline{x}_1]}$	$\dfrac{\wp[\overline{f} \cdot \overline{x}_1 \cdot \overline{x}_2]}{\wp[\overline{x}_1]}$
H_3	1	2	$\dfrac{\wp[f \cdot x_1 \cdot x_2]}{\wp[x_1]}$	$\dfrac{\wp[\overline{f} \cdot x_1 \cdot \overline{x}_2]}{\wp[x_1]}$
H_4	2	1	$\dfrac{\wp[f \cdot \overline{x}_1 \cdot \overline{x}_2 \cdot x_3]}{\wp[\overline{x}_1 \cdot \overline{x}_2]}$	$\dfrac{\wp[\overline{f} \cdot \overline{x}_1 \cdot \overline{x}_2 \cdot \overline{x}_3]}{\wp[\overline{x}_1 \cdot \overline{x}_2]}$
H_5	2	1	$\dfrac{\wp[f \cdot \overline{x}_1 \cdot x_2 \cdot x_3]}{\wp[\overline{x}_1 \cdot x_2]}$	$\dfrac{\wp[\overline{f} \cdot \overline{x}_1 \cdot x_2 \cdot \overline{x}_3]}{\wp[\overline{x}_1 \cdot x_2]}$
H_6	2	1	$\dfrac{\wp[f \cdot x_1 \cdot \overline{x}_2 \cdot x_3]}{\wp[x_1 \cdot \overline{x}_2]}$	$\dfrac{\wp[\overline{f} \cdot x_1 \cdot \overline{x}_2 \cdot \overline{x}_3]}{\wp[x_1 \cdot \overline{x}_2]}$
H_7	2	1	$\dfrac{\wp[f \cdot x_1 \cdot x_2 \cdot x_3]}{\wp[x_1 \cdot x_2]}$	$\dfrac{\wp[\overline{f} \cdot x_1 \cdot x_2 \cdot \overline{x}_3]}{\wp[x_1 \cdot x_2]}$

$$H_k = 2^{n-i}[2(p_{m0} + p_{m1}) - 1] \tag{8.11}$$

The result of Equation 8.8 reduces the computation of a single normalized Haar spectral coefficient to that of finding matching percentages of identical similar outputs of f and a transformation matrix row-function. This can be accomplished by applying the output probability computation algorithm to an OBDD representation of the f_c functions. Using the result of Bayes' theorem, the cofactor output probabilities can be computed by ANDing various cubes with the original function f and dividing the result by the output probability of the cube itself, which is a constant.

The following table contains symbols for each of the Haar spectral coefficients (H_i) values that indicate the size of the cofactor function range (i) and probability expressions that evaluate whether the function to be transformed and the row function simultaneously evaluate to logic-0 (denoted as p_{m0}), or evaluate to logic-1 (denoted as p_{m1}).

By observing that the p_{m0} and p_{m1} expressions for a given H_k in Table 8.1 are statistically independent, the individual computations may be combined into a compact form. As an example, consider H_5.

Example 8.2 *The divisor for the p_{m0} and p_{m1} expressions $\wp[\overline{x}_1 x_2]$ is a constant equal to $\frac{1}{2^i}$ and thus may be factored out resulting in Equation 8.11 being rewritten as*

$$H_k = 2^{n-i}[2^{i+1}(p_{m0} + p_{m1}) - 1] \tag{8.12}$$

Since the Boolean expressions $f \cdot \overline{x}_1 \cdot x_2 \cdot x_3$ and $\overline{f} \cdot \overline{x}_1 \cdot x_2 \cdot \overline{x}_3$ are disjoint, the overall probability may be computed as the sum of the individual probabilities, or alternatively, as the probability of the inclusive-OR of the functions. This is true because it is easy to see that $\wp[g + h] = \wp[g] + \wp[h]$ for g and h that are covered by disjoint cube sets.

Combining the Boolean arguments and simplifying

$$f \cdot \overline{x}_1 \cdot x_2 \cdot x_3 + \overline{f} \cdot \overline{x}_1 \cdot x_2 \cdot \overline{x}_3 = \overline{x}_1 x_2 (\overline{x_3 \oplus f}) \tag{8.13}$$

Therefore, we can rewrite Equation 8.12 as

$$H_k = 2^{n-i}[2^{i+1} P[\overline{x}_1 x_2 (\overline{x_3 \oplus f})] - 1] \tag{8.14}$$

The manipulations used in Example 8.2 may be applied to all of the normalized Haar spectrum coefficients. This leads to the interesting result that the normalized Haar coefficients depend on the set of $n + 1$ Boolean relations, $\{\overline{f \oplus 0}, \overline{f \oplus x_1}, \overline{f \oplus x_2}, \cdots, \overline{f \oplus x_n}\}$, which describe the equivalence of a particular dependent variable x_i and the function to be transformed f. We refer to this set of functions as the *characteristic equivalence relations*. Higher ordered coefficients are based on disjoint partitions of the range space of these equivalence functions. The partitioning is accomplished by ANDing the equivalence functions with various cubes of other dependent variables of f referred to as the *characteristic cubes*. The specific cofactor from which p_m is computed is given by the inherent order of the dependent variables describing f.

Table 8.2 contains the probability functions for an $n = 3$ variable transformation in terms of the characteristic equivalence relations. Using this table, each coefficient can be computed using Equations 8.15 and 8.16.

$$H_i = 2^{n-j}[2^{j+1}p_m - 1] \tag{8.15}$$

$$j = \begin{cases} 0, & i = 0 \\ \lfloor log_2(i) \rfloor, & i > 0 \end{cases} \tag{8.16}$$

We can also compute the total number of possible different valued coefficients for a particular i (or equivalently, a particular j). We note that the Haar coefficients range in value as given by:

$$\{-2^{n-j}, -2^{n-j} + 2, -2^{n-j} + 4, \ldots, -2, 0, +2, \ldots, 2^{n-j} - 4, 2^{n-j} - 2, 2^{n-j}\}$$

Thus, the total number of possible different valued coefficients (denoted by N_j) is given in Equation 8.17.

$$N_j = 2^{n-j} + 1 \tag{8.17}$$

Probabilistic Equivalence Checking

By the definition of event E and the assumption that all functions of n variables are equally likely to arise (uniform distribution), it is easy to see that:

$$\wp[E] = \frac{1}{2^{2^n}} \tag{8.18}$$

Since the normalized Haar spectrum for a given fully specified Boolean function is unique [92], Equation 8.19 holds.

$$\wp[S_i|E] = 1 \tag{8.19}$$

Table 8.2   Relationship of the Haar Spectrum and Characteristic Equivalence Functions

Symbol	j	$n - j$	p_m
H_0	0	3	$P[\overline{0} \cdot f \oplus 0]$
H_1	0	3	$P[\overline{0} \cdot \overline{f \oplus x_1}]$
H_2	1	2	$P[\overline{x}_1 \cdot \overline{f \oplus x_2}]$
H_3	1	2	$P[x_1 \cdot \overline{f \oplus x_2}]$
H_4	2	1	$P[\overline{x}_1 \cdot \overline{x}_2 \cdot \overline{f \oplus x_3}]$
H_5	2	1	$P[\overline{x}_1 \cdot x_2 \cdot \overline{f \oplus x_3}]$
H_6	2	1	$P[x_1 \cdot \overline{x}_2 \cdot \overline{f \oplus x_3}]$
H_7	2	1	$P[x_1 \cdot x_2 \cdot \overline{f \oplus x_3}]$

Equation 8.19 may be generalized for the occurrence of any subset of q events, $\{S_i\}$, to that shown in Equation 8.20.

$$\wp\left[\bigcap_{i=1}^{q} Si \,|\, E\right] = 1 \tag{8.20}$$

Also we see that $\wp[S_i]$ is the ratio of all possible functions that yield the coefficient $H_i(f)$ divided by the total population of 2^{2^n}. We define a counting function $k(H_i)$ that is integer valued and yields the number of fully specified Boolean functions for which the ith Haar spectral coefficient is some specific value. Thus we can express this relationship as shown in Equation 8.21.

$$\wp[S_i] = \frac{k(H_i)}{2^{2^n}} \tag{8.21}$$

From probability theory we know that Equation 8.22 holds.

$$\wp[E \bigcap S_i] = \wp[S_i|E]\wp[E] = \wp[E|S_i]\wp[S_i] \qquad (8.22)$$

Using the relationships in Equations 8.22, 8.20 and 8.18, we see that the conditional probability becomes

$$\wp[E|S_i] = \frac{\wp[E]}{\wp[S_i]} = \frac{1}{k(H_i)} \qquad (8.23)$$

In general, for any subset of events, $\{S_i\}$, we have the expression as given in Equation 8.24.

$$\wp[E| \bigcap_{i=1}^{q} S_i] = \frac{\wp[E \bigcap (\bigcap_{i=1}^{q} S_i)]}{\wp[\bigcap_{i=1}^{q} S_i]} = \left(\frac{1}{2^{2^n}}\right)\left(\frac{1}{\wp[\bigcap_{i=1}^{q} S_i]}\right) = \frac{1}{2^{2^n}\,\wp[\bigcap_{i=1}^{q} S_i]}$$

$$(8.24)$$

Equation 8.24 is the governing expression for the probabilistic equivalence checking technique described here. We see that given a subset of matching Haar spectral coefficients for two functions f and g (or alternatively, a subset of events $\{S_i\}$) the probability that f and g are indeed equivalent may be computed. By obtaining the information that a new event S_i has occurred, we may update the value $\wp[\bigcap_{i=1}^{q} S_i]$ thereby increasing the value $\wp[E|\bigcap_{i=1}^{q} S_i]$.

Relation of Haar Coefficients to Probabilistic Events

This section will derive the relationship between the probabilistic events S_i and their dependence upon the corresponding Haar spectral coefficients $H_i(f)$ and $H_i(g)$. The Haar spectral coefficients may be obtained through the use of any efficient method such as those in [65, 83, 152].

We note that given the ith Haar spectral coefficient for a function f and a function g there appear to be four possibilities as given in Table 8.3. It is seen that as soon as $H_i(f) \neq H_i(g)$ occurs, it is possible to declare $f \neq g$ and to terminate the process of equivalence checking. However, when $H_i(f) = H_i(g)$, it is not known whether $f = g$ or $f \neq g$ unless all possible H_i are found to be

Table 8.3 Apparent Possibilities Given f, g, $H_i(f)$ and $H_i(g)$

Function Relation	H_i Relation	Observation
$f = g$	$H_i(f) = H_i(g)$	Possible $f = g$
$f = g$	$H_i(f) \neq H_i(g)$	Not Possible
$f \neq g$	$H_i(f) = H_i(g)$	Possible $f = g$
$f \neq g$	$H_i(f) \neq H_i(g)$	$f \neq g$

equivalent. However, it is possible to successively refine the $\wp[E|\bigcap_{i=1}^{q} S_i]$ value using Equation 8.24.

For this probabilistic scheme to be practically useful, we need to determine the joint distribution, $\wp[\bigcap_i^{j<2^n-1} S_i]$, as a function of the corresponding subset of Haar spectral coefficients. We first consider the simple case of determining a function for $\wp[S_i]$ that depends on the single Haar spectral coefficient H_i. For a single matching coefficient, we are interested in finding $\wp[E|S_i]$. Since it is known that $\wp[E \cap S_i] = \wp[S_i]\wp[E|S_i]$, we can express the conditional probability as given in Equation 8.25 since $\wp[S_i] \neq 0$.

$$\wp[E|S_i] = \frac{\wp[E \cap S_i]}{\wp[S_i]} \tag{8.25}$$

The numerator of Equation 8.25 is the percentage of functions f and g that have a common Haar coefficient H_i. Since all equivalent functions have the same Haar spectra by the uniqueness property of the transform, we see that $\wp[E \cap S_i] = 1/2^{2^n}$. The denominator of Equation 8.25 is the percentage of functions that have a common H_i value. In general, many different functions can have common H_i values. For example, 6 out 16 possible functions of $n = 2$ variables have $H_0 = 0$. Based on the definition of the counting function $k(H_i)$ we can then express $\wp[S_i] = k(H_i)/2^{2^n}$, and Equation 8.25 is rewritten as Equation 8.26.

$$\wp[E|S_i] = \frac{1}{k(H_i)} \tag{8.26}$$

The relationship between the characteristic equivalence functions and the Haar spectral coefficients is established in the following results.

Lemma 8.1 *Two Boolean functions, $f(x_1, x_2, \ldots, x_n)$ and $g(x_1, x_2, \ldots, x_n)$ can not be equivalent if it is true that $\wp[\overline{f \oplus x_i}] \neq \wp[\overline{g \oplus x_i}]$.*

Proof: Assume the contradiction of the lemma, that is $\wp[\overline{f \oplus x_i}] \neq \wp[\overline{g \oplus x_i}]$, but $f = g$. Since $f = g$, then it must be true that $\overline{f \oplus x_i} = \overline{g \oplus x_i}$ and that $\wp[\overline{f \oplus x_i}] = m_f/2^n$ and $\wp[\overline{g \oplus x_i}] = m_g/2^n$ where m_f is the number of distinct 1-values in the truth vector of $\overline{f \oplus x_i}$ and m_g is the number of distinct 1-values in the truth vector of $\overline{g \oplus x_i}$. But since $\wp[f] = \wp[g]$ and $\overline{f \oplus x_i} = \overline{g \oplus x_i}$, it must be the case that $m_f = m_g$, thus contradicting the assumption that $\wp[\overline{f \oplus x_i}] \neq \wp[\overline{g \oplus x_i}]$.   □

Corollary 8.1 *Two cofactors about the same cube of $\overline{f \oplus x_i}$ and $\overline{g \oplus x_i}$ have identical output probabilities if $f = g$.*

We denote f_{ci} as the function that is formed as the intersection of some cube and i^{th} characteristic equivalence function. Thus, f_{ci} depends upon all n variables and $p_m = \wp[f_{ci}]$. The total number of functions with a common H_i value (denoted as $k(H_i)$) can be computed as the total number of different f_{ci} functions that have a common $\wp[f_{ci}]$ due to Corollary 8.1. Therefore, $k(H_i)$ is the different number of ways that a function with a range space of size 2^{n-j} can have $2^n p_m$ logic-1 values. This combinatorial quantity must be scaled by a constant to account for the decreasing magnitude of the H_i that are distributed over the entire population of 2^{2^n} functions as i increases. This scaling factor is seen to be $2^{N_0 - N_j}$ where N_i is defined in Equation 8.17. Given these observations, $k(H_i)$ can be expressed as

$$k(H_i) = 2^{N_0 - N_j} \binom{2^{n-j}}{2^n p_m} \tag{8.27}$$

Using the fact that $2^{N_0 - N_j} = 2^{2^n - 2^{n-j}}$ and that $p_m = \frac{H_i + 2^{n-j}}{2^{n+1}}$, we can reduce Equation 8.27 to Equation 8.28.

$$k(H_i) = 2^{2^n - 2^{n-j}} \begin{pmatrix} 2^{n-j} \\ \frac{H_i + 2^{n-j}}{2} \end{pmatrix} \tag{8.28}$$

Thus, we can rewrite Equation 8.26 as shown in Equation 8.29.

$$\wp[E|S_i] = \left(\frac{1}{2^{2^n - 2^{n-j}}} \right) \begin{pmatrix} 2^{n-j} \\ \frac{H_i + 2^{n-j}}{2} \end{pmatrix}^{-1} \tag{8.29}$$

As an example, consider the case where $H_0(f) = H_0(g) = 2$ for $n = 2$ variables. To compute $\wp[E|S_0]$, we use the relationship in Equation 8.29 resulting in $\wp[E|S_0] = \frac{1}{4}$.

To successively improve the $\wp[E|S_i]$ value, it must be updated with each subsequent event S_i. For a practical implementation, this means that $\wp[\bigcap_i^{j < 2^n - 1} S_i]$ must be computed as a function of the corresponding Haar coefficients H_i. However, this value cannot be computed as a simple product of individual $\wp[S_i]$ values since the multiple S_i events are not necessarily statistically independent.

For the general case, we must re-compute $\wp[\bigcap_{i=1}^q S_i]$ for each new event S_i in order to update the $\wp[E| \bigcap_{i=1}^q S_i]$ value. Some events are statistically dependent while other subsets are not. Recall that the values H_i depend on cofactor functions of various characteristic equivalence functions about some cube. A subset of events $\{S_i\}$ are all statistically independent if they result from a corresponding subset of matching Haar spectral coefficients $\{H_i\}$ that are formed based on Shannon cofactors with respect to mutually disjoint cubes. As an example, H_2 and H_3 result from the functions $\overline{x}_1 \cdot f \oplus x_2$ and $x_1 \cdot f \oplus x_2$ which are disjoint. Thus, $\wp[S_2 \cap S_3] = \wp[S_2]\wp[S_3]$ since $\overline{x}_1$ and x_1 are disjoint characteristic cubes.

Not all events $\{S_i\}$ are statistically independent. As an example, H_0 and H_1 are dependent since an intersection of the cofactors of the characteristic equivalence functions of H_0 and H_1 exists and is non-null. In order to find the value $\wp[S_1 \cap S_0]$, we generalize our definition of the counting function to $k(H_0, H_1)$ which will denote the number of possible Boolean functions that may have both H_0 and H_1 as Haar spectral coefficients. Given this quantity, we may then express the desired joint probability as given in Equation 8.30.

$$\wp[S_0 \bigcap S_1] = \frac{k(H_0, H_1)}{2^{2^n}} \tag{8.30}$$

In general, we have Equation 8.31 resulting in Equation 8.32.

$$\wp[\bigcap_{i=0}^{q} S_i] = \frac{k(H_i, H_{i+1}, \ldots, H_q)}{2^{2^n}} \tag{8.31}$$

$$\wp[E| \bigcap_{i=0}^{q} S_i] = \frac{1}{2^{2^n} \wp[\bigcap_{i=1}^{q} S_i]} = \frac{1}{k(H_i, H_{i+1}, \ldots, H_q)} \tag{8.32}$$

To compute this joint probability distribution, we must have some information concerning the dependent relationship between individual $k(H_i)$ and $k(H_m)$ values. As an example of this dependence, we will derive the relationship between H_0 and H_1. For $i = 0, 1$, the integer j in Table 8.2 is zero valued yielding the relationships as shown in Equations 8.33 and 8.34.

$$H_0 = 2^n[2p_{m0} - 1] \tag{8.33}$$

$$H_1 = 2^n[2p_{m1} - 1] \tag{8.34}$$

For these two coefficients, we have the output probabilities p_{m0} and p_{m1} that may be expressed as:

$$p_{m0} = \wp[\overline{0} \cdot \overline{f \oplus 0}] = \wp[\overline{f}] = \frac{1}{2}\{\wp[\overline{f}_{\overline{x}_1}] + \wp[\overline{f}_{x_1}]\} \tag{8.35}$$

$$p_{m1} = \wp[\overline{0} \cdot \overline{f \oplus x_1}] = \wp[\overline{f \oplus x_1}] = \frac{1}{2}\{\wp[f_{x_1}] + \wp[\overline{f}_{\overline{x}_1}]\} \tag{8.36}$$

Thus, the corresponding Haar spectral coefficients become

$$H_0 = 2^n \{ \wp[\vec{f}_{\overline{x}_1}] + \wp[\vec{f}_{x_1}] - 1 \} \tag{8.37}$$

$$H_1 = 2^n \{ \wp[\overleftarrow{f}_{\overline{x}_1}] + \wp[f_{x_1}] - 1 \} \tag{8.38}$$

Equating the $\wp[\overleftarrow{f}_{\overline{x}_1}]$ values results in the relationship between H_0 and H_1 as given in Equation 8.39.

$$H_1 = H_0 + 2^n \{ \wp[f_{x_1}] - \wp[\overline{f}_{x_1}] \} \tag{8.39}$$

Using the result of Equations 8.28 and 8.39, we have

$$k(H_1) = \left(\begin{array}{c} 2^n \\ \frac{H_1 + 2^n}{2} \end{array} \right) = \left(\begin{array}{c} 2^n \\ \frac{H_0 + 2^n \{ \wp[f_{x_1}] - \wp[\overline{f}_{x_1}] \} + 2^n}{2} \end{array} \right) = \left(\begin{array}{c} 2^n \\ \frac{H_0 + 2^n}{2} + \frac{H_1 - H_0}{2} \end{array} \right) \tag{8.40}$$

Using the identity relation

$$\left(\begin{array}{c} n \\ k \end{array} \right) \left(\begin{array}{c} n - k \\ m - k \end{array} \right) = \left(\begin{array}{c} n \\ m \end{array} \right) \left(\begin{array}{c} m \\ k \end{array} \right) \tag{8.41}$$

We rewrite Equation 8.40 as

$$\left(\begin{array}{c} 2^n \\ \frac{H_0 + 2^n}{2} \end{array} \right) \left(\begin{array}{c} \frac{H_0 + 2^n}{2} \\ \frac{H_1 + 2^n}{2} \end{array} \right) = \left(\begin{array}{c} 2^n - \frac{H_1 + 2^n}{2} \\ \frac{H_0 + 2^n}{2} - \frac{H_1 + 2^n}{2} \end{array} \right) \left(\begin{array}{c} 2^n \\ \frac{H_1 + 2^n}{2} \end{array} \right) \tag{8.42}$$

Equation 8.42 reduces to Equation 8.43, and we define the resulting quantity as the value $A_{0,1}$.

$$A_{0,1} = \frac{k(H_0)}{k(H_1)} = \left(\frac{2^n - \frac{H_1+2^n}{2}}{\frac{H_0+2^n}{2} - \frac{H_1+2^n}{2}} \right) \left(\frac{\frac{H_0+2^n}{2}}{\frac{H_1+2^n}{2}} \right)^{-1} \tag{8.43}$$

Thus, we have the result that the values $k(H_0)$ and $k(H_1)$ are deterministically related as $k(H_0) = A_{0,1}k(H_1)$. Using this fact, we devise a means for computing the desired joint quantity $k(H_0, H_1)$. We know that for a given H_0 and H_1 to exist for a single function, the corresponding $k(H_0) - A_{0,1}k(H_1) = 0$. Thus, it is possible to check all possible $k(H_1)$ values for $H_1 = \{-2^n, -2^n + 2, \ldots, -2, 0, 2, \ldots, 2^n - 2, 2^n\}$ and where the relationship is satisfied, we increment the value of $k(H_0, H_1)$. This may be expressed in closed form through the use of the unit-step function $u(\tau)$ as defined in Equation 8.44.

$$u(\tau) = \begin{cases} 1, & \tau = 0 \\ 0, & otherwise \end{cases} \tag{8.44}$$

Equation 8.45 then expresses the desired relationship.

$$k(H_0, H_1) = \sum_{i=0}^{2^n+1} u[k(H_0) - A_{0,1}k(H_i)] \tag{8.45}$$

The upper bound of the summation is the total number of possible Haar spectral coefficients that can result for the $i = 1$ coefficient. Although the complexity of this approach is prohibitive for low-ordered coefficients, Equation 8.45 can be algorithmically stated as follows:

1. Compute $A_{0,1}$ based on the value of the Haar spectral coefficients.

2. Initialize $k(H_0, H_1) = 0$.

3. Loop over all possible $2^n + 1$ values of H_1 and evaluate if $(k(H_0) - A_{0,1}k(H_1)) = 0$, then increment the value of $k(H_0, H_1)$.

4. $P[S_0 \cap S_1] = \frac{k(H_0,H_1)}{2^{2^n}}$, or, $P[E|(S_0 \cap S_1)] = \frac{1}{k(H_0,H_1)}$.

This procedure can be generalized by deriving and computing new $A_{i,j,\ldots,k}$ values each time a new event S_k occurs and updating the $P[\bigcap_{i=1}^{q} S_i]$ and hence, the $P[E|\bigcap_{i=1}^{q} S_i]$ values.

In general, this procedure leads to exponential complexity for updating all new k values. However, we note that the multi-resolution nature of the Haar transform allows us to determine subsets of coefficients that are statistically independent, thus avoiding the computation of large joint distributions. This can also be coupled with the construction of the partial BDDs by constraining them to represent mutually disjoint portions of the functions under consideration. We also note that the decreased range space dimension of high-ordered coefficients H_i can allow the algorithm to run in reasonable time for those H_i.

8.3.3   Example Calculation

As an example, consider Table 8.4 which contains the Haar spectral vectors for all possible functions of $n = 2$ variables. We will assume that we are dealing with two functions, $f(x_1, x_2)$ and $g(x_1, x_2)$, such that f and g are equivalent to the function represented in the third row of Table 8.4. Thus, the corresponding Haar spectral vector is $\vec{H}^T(f) = \vec{H}^T(g) = (H_0, H_1, H_2, H_3) = (2, 2, 0, -2)$. Figure 8.8 contains the Karnaugh maps and corresponding partial and complete BDDs for the function f (or g). Note that the BDDs are also interpreted as HSDs with the $1 - edges$ having an attribute equal to a Haar spectral coefficient value. The coefficient attributes are shown on the HSD/BDDs with an "*" indicating that the exact coefficient could not be computed. From the center partial HSD/BDD, we see that $H_2 = 0$, and from the rightmost partial HSD/BDD we see that $H_3 = -2$.

From the partial BDDs it is seen that only two Haar spectral coefficients can be obtained, H_2 and H_3. This is due to the fact that H_0 and H_1 require a completely specified HSD since the corresponding transform matrix rows have no 0-valued entries. For more practical cases with much larger values of n, we obtain a larger fraction of the total number of Haar coefficients than the 50% obtained from this small example.

Table 8.4   All Possible Boolean Functions for $n = 2$ and their Haar Spectra

Function, f				H_0	H_1	H_2	H_3	Expression
0	0	0	0	4	0	0	0	0
0	0	0	1	2	-2	0	2	xy
0	0	1	0	2	2	0	-2	$x\overline{y}$
0	0	1	1	0	4	0	0	x
0	1	0	0	2	-2	2	0	$\overline{x}y$
0	1	0	1	0	0	2	2	y
0	1	1	0	0	0	2	-2	$x \oplus y$
0	1	1	1	-2	2	2	0	$x + y$
1	0	0	0	2	-2	-2	0	$\overline{xy}$
1	0	0	1	0	0	-2	2	$\overline{x \oplus y}$
1	0	1	0	0	0	-2	-2	$\overline{y}$
1	0	1	1	-2	2	-2	0	$x + \overline{y}$
1	1	0	0	0	-4	0	0	$\overline{x}$
1	1	0	1	-2	-2	0	2	$\overline{x} + y$
1	1	1	0	-2	-2	0	-2	$\overline{x} + \overline{y}$
1	1	1	1	-4	0	0	0	1

Using the previously derived equations, we have $k(H_2) = 8$, $k(H_3) = 4$ and $k(H_2, H_3) = 2$. These values result in the probability values $\wp[E|S_2] = \frac{1}{8}$, $\wp[E|S_3] = \frac{1}{4}$ and $\wp[E|(S_2 \cap S_3)] = \frac{1}{2}$. Furthermore, we note that $\wp[S_2 \cap S_3] = \wp[S_2]\wp[S_3] = (\frac{1}{2})(\frac{1}{4}) = \frac{1}{8}$ in this case since S_2 and S_3 are statistically independent. The independence arose from the fact that the two partial BDDs represent disjoint segments of the range space of the function. If this is ensured during the construction of all partial BDDs, the joint computation of $k(H_2, H_3)$ may be avoided and $P[E|(S_2 \cap S_3)]$ may be computed as given in Equation 8.46.

$$\wp[E|(S_2 \cap S_3)] = \frac{1}{2^{2n}\,\wp[S_2 \cap S_3]} = \frac{1}{2^{2n}\,\wp[S_2]\wp[S_3]} = \frac{1}{16\frac{1}{8}} = \frac{1}{2} \qquad (8.46)$$

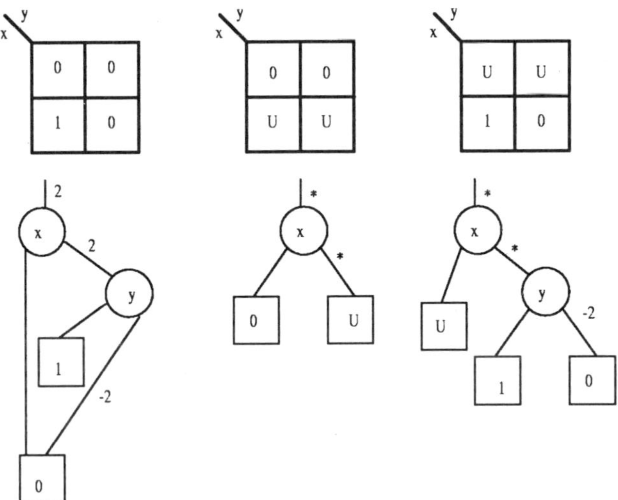

Figure 8.8   Karnaugh Maps and HSD/BDDs of Complete and Partial Functions

This result shows that there are only 2 possible functions out of the population of $2^{2^n} = 16$ that have $H_2 = 0$ and $H_3 = -2$.

8.3.4   Experimental Results

Experiments were formulated to investigate the effectiveness of using Haar spectral coefficients for equivalence checking. These experiments were run to observe the average number of Haar coefficients needed before a mismatch in value was found for two functions known to be slightly different. The results also give an indication of how different errors between two versions of a circuit affect the number of required Haar coefficients for a mismatch to be found.

The initial set-up for this experiment involved choosing a single output from a benchmark function and randomly inserting a single inverter in the netlist. Next, HSDs were formed for the circuit with the inverter and without. To ensure the two HSDs did indeed represent different functions, a graphical equivalence checker was used. The experiment consisted of randomly extracting pairs of same-order Haar coefficients from the two representations until two were found that differed in value. For each given circuit error (that is, each given inverter insertion) 1024 trials were made.

Table 8.5 Effect of Different Errors on Haar Coefficient Matching

Circuit	Inp	Inverter Error (5 Random Trials)					
c432	36	avg	58.0	24.8	474.1	15.9	58.3
		dev	61.3	24.5	471.2	15.4	55.3
c499	41	avg	69.3	65.3	61.4	68.3	66.1
		dev	70.4	66.6	60.1	66.9	65.3
c880	42	avg	669.3	393.4	36.6	160.7	768.6
		dev	681.1	385.9	36.7	161.4	763.9
c2670	78	avg	29.9	9.2	10.5	128.1	7.5
		dev	28.5	9.3	10.0	129.0	7.2
cm151a	12	avg	5.3	16.1	6.4	6.5	15.9
		dev	4.7	16.3	5.9	6.1	15.8
cu	13	avg	22.5	16.1	48.5	80.2	162.6
		dev	22.2	15.9	47.6	85.0	162.1
misex3	14	avg	326.6	28.1	579.5	337.1	303.1
		dev	336.0	27.1	576.0	324.3	287.3
frg1	25	avg	39866.3	70.1	471.6	104225.8	1709.9
		dev	38715.4	70.9	456.9	104031.3	1740.4
too_large	36	avg	636.5	2559.8	104685.6	1169.4	640.6
		dev	616.7	2667.7	103407.8	1176.7	609.8
t481	16	avg	810.6	503.9	415.2	53.3	3.1
		dev	760.4	508.1	425.4	52.0	2.6

Table 8.5 contains the results for 10 benchmark functions, each with 10 different inverter errors. The column labeled *Inp* contains the number of distinct variables in the dependence set of the function, and the row labeled *avg* is the average number of Haar coefficients (over the 1024 trials) that were required before a mismatch occurred. Likewise, the row labeled *dev* contains the standard deviation of the number of required Haar coefficients. It is apparent that the standard deviation is approximately the same value as the mean in all cases. This is a result of the fact that the subset of Haar coefficients was chosen randomly with the assumption that each was equally likely for two designs that are known to differ (*i.e.*, a geometric distribution resulted in terms of the average number of coefficients before a mismatch occurred). Although this observation is largely an artifact of our experimental setup, another result is the large range in value of the required number of coefficients in order to detect the differences in the two circuits. As an example, we see that the benchmark *frg1* has differences in the averages that are as great as four orders of magnitude (eg. 70.1 versus 104225.8).

The data presented in Table 8.6 was computed in order to compare the Haar coefficient matching scheme to random simulations. These results compare the average number of required Haar coefficients to the number of random simulations that must be performed before a difference in the two circuits is detected. The simulations were performed using equally likely, randomly generated test vectors. The averages were formed over the 10 circuit modifications described above with 1024 trials each. In terms of comparing just the number of simulations to required Haar coefficients, we see that each technique is approximately equal since of the 21 benchmark functions in Table 8.6, 13 required fewer coefficients than random simulations.

However, we must point out the very important fact that the computational overhead required to obtain a single Haar coefficient is not equal to that for performing a simulation. Furthermore, the assumption that the Haar coefficients are equally likely to occur also biases these results to some degree since the subset of coefficients resulting from a specific partial HSD will have mutual dependence due to the definition of the transform. Nevertheless, we can conclude that the use of Haar coefficients does appear to yield as much information as random simulations over this sample of benchmark functions. The importance of this result is that schemes that allow for the computation of Haar coefficients more efficiently than a netlist simulation can be used to increase the effectiveness of statistical verification.

8.4 SUMMARY

In this section logic verification has been studied. After briefly reviewing the standard techniques a method for probabilistically determining the equivalence of two Boolean functions has been developed and presented. We have combined the use of two notions: partial BDDs [134] and the computation of Haar spectral coefficients using a BDD as an HSD [83]. The probabilistic framework has been derived for the equivalence checking problem.

Experimental results indicate that this approach can be a viable alternative for equivalence checking of functions that are difficult to represent completely. The experiments also indicate that this approach can be better in terms of required computational resources as compared to a repeated simulation approach.

Table 8.6 Average Number of Haar Coefficients Before a Mismatch Occurs

Circuit	Inp	Avg Number Coefficients	Avg Number Simulations
9sym-hdl	*9*	*2.7*	*5.9*
c2670.329	78	39.9	29.1
c432.432GAT	36	81.1	43.2
c499.OD31	*41*	*74.2*	*252.4*
c880.880GAT	42	242.9	119.1
cc.10	7	16.0	3.8
cm150a	21	7373.9	55.1
cm151a.m	12	54.8	22.1
cm162a.r	11	33.5	27.3
cu.v	*13*	*103.3*	*223.5*
dalu.O7	*57*	*3133.2*	*3584.3*
frg1.d0	*25*	*20349.2*	*26958.6*
misex3.l2	*14*	*212.1*	*1586.0*
mux	21	29810.1	93.1
pcler8.q0	*13*	*641.4*	*1567.7*
pm1.c0	*9*	*25.4*	*90.5*
rd53-hdl.out<2>	*5*	*6.8*	*9.8*
t481	*16*	*443.0*	*2214.2*
too_large.n0	*36*	*12094.8*	*94946.1*
x2.p	*10*	*15.9*	*31.3*
z4ml.24	*7*	*7.4*	*32.1*

9

CONCLUDING REMARKS

In this book, we have explored spectral techniques with a view to applications in VLSI CAD. In particular, it has been shown that the use of decision diagrams allows spectral techniques, which previously have been computationally quite limited, to be applied to realistic problems.

Chapter 3 reviewed the Walsh, Reed-Muller, arithmetic and Haar spectral domains and investigated the transformation amongst those domains as well as transformation to and from the Boolean domain. Chapter 4 covered the fundamentals of a variety of decision diagram types and provides the reader with both the necessary background on those types and the breadth of understanding required to explore other decision diagram types that might arise.

Chapter 5 considered the critical issue of the efficient spectral computation. Cube based techniques were reviewed. The Cooley-Tukey fast computation methods were transformed to graph algorithms suitable for implementation using decision diagrams. Computation of partial spectra was also considered. The techniques presented provide for efficient spectral computation and also represent a basis for further developments in this area.

Variable order is a critical factor in the size of a decision diagram whether it is used to represent a function in the Boolean or in a spectral domain. In Chapter 6, a variable ordering approach was presented that is computationally similar to the well-known sifting approach but which generally yields smaller BDDs. The important area of linear transformation in relation to variable reordering in BDDs was extensively considered in Chapter 6. Some functions show exponential BDD size reduction when linear transformations are applied and the exponential blow-up encountered by other variable ordering techniques is avoided. A window optimization technique was also presented that with small computational cost can significantly improve the results. This work demonstrates the power of these various approaches and should lead to further work in the development of both heuristic and exact methods for improving BDD size. Developments in this area continue to be of importance as BDD methods are applied to larger and larger problems.

Chapter 7 focuses on the application of spectral methods to logic synthesis. The topics considered included: spectral translation, the maximum correlation iterative approach, synthesis using heuristic techniques and spectral methods for technology-mapping. The reader should take these as representative of the application of spectral methods to synthesis and certainly not exhaustive. Our goal has been to show that, largely due to the use of decision diagram techniques, spectral methods now indeed do deserve consideration in developing CAD methods for VLSI design.

Finally, Chapter 8 addressed logic verification and, in particular, introduced a method employing the Haar spectral domain for probabilistically determining the equivalence of two Boolean functions. The results presented demonstrated this approach can be computationally better than alternative methods.

The areas of decision diagrams and spectral methods are very broad, rapidly developing ones and this book has of necessity focused on key fundamentals and representative developments. We hope the work presented does show the potential benefit of spectral approaches and that it will encourage readers to explore the extensive and fruitful literature and in addition encourage some readers to direct their research efforts to further developments.

REFERENCES

[1] N. Ahmed and K. R. Rao. *Orthogonal Transforms for Digital Signal Processing.* Springer-Verlag, New York, New York, 1975.

[2] S.B. Akers. Binary decision diagrams. *IEEE Trans. on Comp.*, 27:509–516, 1978.

[3] P. Ashar, S. Devadas, and K. Keutzer. Path-delay-fault testability properties of multiplexor-based networks. *INTEGRATION, the VLSI Jour.*, 15(1):1–23, 1993.

[4] R.I. Bahar, E.A. Frohm, C.M. Gaona, G.D. Hachtel, E. Macii, A. Pardo, and F. Somenzi. Algebraic decision diagrams and their application. In *Int'l Conf. on CAD*, pages 188–191, 1993.

[5] K. G. Beauchamp. *Applications of Walsh and Related Functions.* Academic Press, 1984.

[6] B. Becker. Synthesis for testability: Binary decision diagrams. In *STACS*, volume 577 of *LNCS*, pages 501–512. Springer Verlag, 1992.

[7] B. Becker, R. Drechsler, and R. Enders. On the computational power of bit-level and word-level decision diagrams. In *ASP Design Automation Conf.*, pages 461–467, 1997.

[8] A. Bernasconi and B. Codenotti. Spectral analysis of Boolean functions as a graph eigenvalue problem. *IEEE Trans. on Comp.*, 48:345–351, 1999.

[9] V. Bertacco, S. Minato, P. Verplaetse, L. Benini, and G. De Micheli. Decision diagrams and pass transistor logic synthesis. In *Int'l Workshop on Logic Synth.*, 1997.

[10] P. W. Besslich. *Spectral Techniques and Fault Detection, (M. G. Karpovsky, editor).* Academic Press Publishers, Boston, Massachusetts, 1985.

[11] B. Bollig and I. Wegener. Improving the variable ordering of OBDDs is NP-complete. *IEEE Trans. on Comp.*, 45(9):993–1002, 1996.

[12] K.S. Brace, R.L. Rudell, and R.E. Bryant. Efficient implementation of a BDD package. In *Design Automation Conf.*, pages 40–45, 1990.

[13] R.K. Brayton, G.D. Hachtel, C. McMullen, and A.L. Sangiovanni-Vincentelli. *Logic Minimization Algorithms for VLSI Synthesis*. Kluwer Academic Publishers, 1984.

[14] R.K. Brayton, R. Rudell, A.L. Sangiovanni-Vincentelli, and A.R. Wang. MIS: A multiple - level logic optimization system. *IEEE Trans. on Comp.*, 6(6):1062–1081, 1987.

[15] R.E. Bryant. Graph-based algorithms for Boolean function manipulation. *IEEE Trans. on Comp.*, 35(8):677–691, 1986.

[16] R.E. Bryant. On the complexity of VLSI implementations and graph representations of Boolean functions with application to integer multiplication. *IEEE Trans. on Comp.*, 40:205–213, 1991.

[17] R.E. Bryant. Symbolic Boolean manipulation with ordered binary decision diagrams. *ACM, Comp. Surveys*, 24:293–318, 1992.

[18] R.E. Bryant and Y.-A. Chen. Verification of arithmetic functions with binary moment diagrams. Technical report, CMU-CS-94-160, 1994.

[19] R.E. Bryant and Y.-A. Chen. Verification of arithmetic functions with binary moment diagrams. In *Design Automation Conf.*, pages 535–541, 1995.

[20] P. Buch, A. Narayan, A.R. Newton, and A.L. Sangiovanni-Vincentelli. Logic synthesis for large pass transistor circuits. In *Int'l Conf. on CAD*, pages 663–670, 1997.

[21] J. R. Burch and D. E. Long. Efficient Boolean function matching. *IEEE Trans. on CAD*, pages 408–411, 1992.

[22] A. Cayley. Desiderata and suggestions. nr. 2: The theory of groups, graphical representation. *American Journal of Mathematics*, 1:171–176, 1878.

[23] A. Cayley. On the theory of groups. *Proceedings of the London Mathematical Society*, 9:126–133, 1878.

[24] S. Chakravarty. On the complexity of using BDDs for the synthesis and analysis of Boolean circuits. In *Allerton Conference on Communication, Control, and Computing*, pages 730–739, Monticello, Illinois, 1989.

[25] K.-C. Chen. Boolean matching based on Boolean unification. In *European Design & Test Conf.*, pages 346–351, 1993.

[26] D.I. Cheng and M. Marek-Sadowska. Verifying equivalence of functions with unknown input correspondence. In *European Conf. on Design Automation*, pages 81–85, 1993.

[27] C. K. Chow. On the characterization of threshold functions. *IEEE Special Publication S.134*, pages 34–38, 1961.

[28] P. Y. Chung, I. N. Hajj, and J. Patel. Efficient variable ordering heuristics for shared ROBDD. In *Int'l Symp. Circ. and Systems*, pages 1690–1693, Chicago, Illinois, 1993.

[29] E. Clarke, M. Fujita, P. McGeer, K.L. McMillan, J. Yang, and X. Zhao. Multi terminal binary decision diagrams: An efficient data structure for matrix representation. In *Int'l Workshop on Logic Synth.*, pages P6a:1–15, 1993.

[30] E. M. Clarke, M. Fujita, and X. Zhao. Multi-terminal binary decision diagrams and hybrid decision diagrams. In T. Sasao and M. Fujita, editors, *Representation of Discrete Functions*, pages 93–108. Kluwer Academic Publishers, 1996.

[31] E. M. Clarke and R. P. Kurshan. Computer-aided verification. *IEEE Spectrum*, pages 61–67, June, 1996.

[32] E. M. Clarke, K. L. McMillan, X. Zhao, M. Fujita, P. McGeer, and J. Yang. Spectral transforms for large Boolean functions with applications to technology mapping. In *Int'l Workshop on Logic Synth.*, pages 6b1–6b15, 1993.

[33] E.M. Clarke, M. Fujita, and X. Zhao. Hybrid decision diagrams - overcoming the limitations of MTBDDs and BMDs. In *Int'l Conf. on CAD*, pages 159–163, 1995.

[34] E.M. Clarke, K.L. McMillan, X. Zhao, M. Fujita, and J. Yang. Spectral transforms for large Boolean functions with application to technology mapping. In *Design Automation Conf.*, pages 54–60, 1993.

[35] J. W. Cooley and J. W. Tukey. An algorithm for the machine calculation of complex Fourier series. *Math. Computation*, 19:297–301, 1965.

[36] D. M. Cvetkovic, M. Doob, and H. Sachs. *Spectra of Graphs*. Academic Press, 1979.

[37] T. Damarla. Generalized transforms for multiple valued circuits and their fault detection. *IEEE Trans. on Comp.*, C-41, no. 9:1101–1109, 1992.

[38] M. Davio, J.P. Deschamps, and A. Thayse. *Discrete and Switching Functions*. McGraw-Hill, 1978.

[39] R. C. Debnath and A. K. Karmarker. Method for finding Rademacher-Walsh spectral coefficients of Boolean functions. *International Journal of Electronics*, 60:245–250, 1986.

[40] E. Detjens, G. Ganot, R. Rudell, and A. Sangiovani-Vincentelli. Spectral transforms for large Boolean functions with applications to technology mapping. In *Int'l Conf. on CAD*, pages 1062–1081, 1987.

[41] R. Drechsler. *Formal Verification of Circuits*. Kluwer Academic Publishers, 2000.

[42] R. Drechsler and B. Becker. – Ordered Kronecker functional decision diagrams – a data structure for representation and manipulation of Boolean functions. *IEEE Trans. on CAD*, 17(10):965–973, 1998.

[43] R. Drechsler and B. Becker. *Binary Decision Diagrams - Theory and Implementation*. Kluwer Academic Publishers, 1998.

[44] R. Drechsler, B. Becker, and N. Göckel. A genetic algorithm for variable ordering of OBDDs. *IEE Proceedings*, 143(6):364–368, 1996.

[45] R. Drechsler, B. Becker, and S. Ruppertz. K*BMDs: A new data structure for verification. In *European Design & Test Conf.*, pages 2–8, 1996.

[46] R. Drechsler, B. Becker, and S. Ruppertz. The K*BMD: A verification data structure. *IEEE Design & Test of Comp.*, pages 51–59, 1997.

[47] R. Drechsler, B. Becker, and S. Ruppertz. Manipulation algorithms for K*BMDs. In *Tools and Algorithms for the Constuction and Analysis of Systems, LNCS*, pages 4–18, 1997.

[48] R. Drechsler, N. Drechsler, and W. Günther. Fast exact minimization of BDDs. In *Design Automation Conf.*, pages 200–205, 1998.

[49] R. Drechsler, N. Drechsler, and W. Günther. Fast exact minimization of BDDs. *IEEE Trans. on CAD*, 19(3):384–389, 2000.

[50] R. Drechsler, A. Hett, and B. Becker. Symbolic simulation using decision diagrams. *Electronic Letters*, 33(8):665–667, 1997.

[51] R. Drechsler, A. Sarabi, M. Theobald, B. Becker, and M.A. Perkowski. Efficient representation and manipulation of switching functions based on ordered Kronecker functional decision diagrams. In *Design Automation Conf.*, pages 415–419, 1994.

[52] R. Drechsler, M. Theobald, and B. Becker. Fast OFDD based minimization of fixed polarity Reed-Muller expressions. In *European Design Automation Conf.*, pages 2–7, 1994.

[53] D. K. Pradhan (editor). *Fault-Tolerant Computing Theory and Techniques Volume I.* Prentice-Hall, Englewood Cliffs, N.J., 1986.

[54] C. R. Edwards. The application of the Rademacher-Walsh transform to Boolean function classification and threshold logic synthesis. *IEEE Trans. on Comp.*, pages 48–62, 1975.

[55] C. R. Edwards. The design of easily tested circuits using mapping and spectral techniques. *Radio and Electronic Engineer*, 47, no. 7:321–342, 1977.

[56] S. Ercolani and G.D. Micheli. Technology mapping for electrically programmable gate arrays. In *Design Automation Conf.*, pages 234–239, 1991.

[57] E. Erolani and G. De Michelli. Technology mapping for electrically programmable gate arrays. In *Design Automation Conf.*, pages 234–239, 1991.

[58] B. Falkowski and I. Schäfer amd M. A. Perkowski. Effective computer methods for the calculation of Rademacher-Walsh spectrum for completely and incompletely specified Boolean functions. *IEEE Trans. on CAD*, 11:1207–1226, 1992.

[59] B. Falkowski and C. Chang. Efficient algorithms for the calculation of arithmetic spectrum from OBDD and synthesis of OBDD from arithmetic spectrum for incompletely specified Boolean functions. In *Int'l Symp. Circ. and Systems*, 1994.

[60] B. Falkowski and C.-H. Chang. Efficient calculation of Gray code-ordered Walsh spectra through algebraic decision diagrams. *Electronic Letters*, 34(9):848–850, 1998.

[61] B. Falkowski and M. A. Perkowski. One more method for the calculation of Hadamard-Walsh spectrum for completely and incompletely specified Boolean functions. *International Journal of Electronics*, 69:595–602, 1990.

[62] B. J. Falkowski and S. Kannurao. Disjoint bi-decompositions of Boolean functions in the Walsh spectral domain. In *The Ninth International Workshop on Post-binary Ultra-large-scale Integration Systems*, pages 15–19, 2000.

[63] B. J. Falkowski and M. A. Perkowski. A family of all essential radix-2 addition/subtraction multi-polarity transforms: Algorithms and interpretations in Boolean domain. In *Int'l Symp. Circ. and Systems*, pages 2913–2916, 1990.

[64] B.J. Falkowski. Properties and ways of calculation of multi-polarity generalized Walsh transforms. *IEEE Trans. on Comp.*, 41:380–391, 1994.

[65] B.J. Falkowski and C.-C. Chang. Efficient algorithms for forward and inverse transformations between Haar spectrum and binary decision diagram. In *International Phoenix Conference on Computers and Communications*, pages 497–503, 1994.

[66] B. J. Fino and V. R. Algazi. A unified treatment of discrete fast transforms. *SIAM Journ. Comput.*, 6:700–717, 1977.

[67] S.J. Friedman and K.J. Supowit. Finding the optimal variable ordering for binary decision diagrams. In *Design Automation Conf.*, pages 348–356, 1987.

[68] H. Fujii, G. Ootomo, and C. Hori. Interleaving based variable ordering methods for ordered binary decision diagrams. In *Int'l Conf. on CAD*, pages 38–41, 1993.

[69] M. Fujita, H. Fujisawa, and N. Kawato. Evaluation and improvements of Boolean comparison method based on binary decision diagrams. In *Int'l Conf. on CAD*, pages 2–5, 1988.

[70] M. Fujita, Y. Matsunaga, and T. Kakuda. On variable ordering of binary decision diagrams for the application of multi-level synthesis. In *European Conf. on Design Automation*, pages 50–54, 1991.

[71] M. Fujita, J. C.-Y. Yang, E. M. Clarke, X. Zhao, and P. McGeer. Fast spectrum computation for logic functions using binary decision diagrams. In *Int'l Symp. Circ. and Systems*, pages 275–278, 1995.

[72] M. Ganai, A. Aziz, and A. Kuehlmann. Enhancing simulation with BDDs and ATPG. In *Design Automation Conf.*, pages 385–390, 1999.

[73] A. Ghosh, S. Devadas, K. Keutzer, and J. White. Estimation of average switching activity in combinational and sequential circuits. In *Design Automation Conf.*, pages 253–259, 1992.

[74] A. Graham. *Kronecker Products and Matrix Calculus: with Applications*. Ellis Horwood Limited and John Wiley & Sons, New York, 1981.

[75] D. Green. *Modern Logic Design*. Addison-Wesley, Reading, Massachusetts, 1986.

[76] W. Günther and R. Drechsler. BDD minimization by linear transformations. In *Advanced Computer Systems*, pages 525–532, 1998.

[77] W. Günther and R. Drechsler. Linear transformations and exact minimization of BDDs. In *Great Lakes Symp. VLSI*, pages 325–330, 1998.

[78] W. Günther and R. Drechsler. On the computational power of linearly transformed BDDs. *Information Processing Letters*, 75(3):119–125, 2000.

[79] A. Gupta. Formal hardware verification methods. A survey. *Formal Methods in System Design*, 1:151–238, 1992.

[80] A. Haar. Zur Theorie der orthogonalen Funktionensysteme. *Math. Ann.*, 69:331–371, 1910.

[81] G.D. Hachtel, E. Macii, A. Pardo, and F. Somenzi. Markovian analysis of large finite state machines. *IEEE Trans. on CAD*, 15(12):1479–1493, 1996.

[82] J. P. Hansen and M. Sekine. Synthesis by spectral translation using Boolean decision diagrams. In *Design Automation Conf.*, pages 248–253, June 1996.

[83] J.P. Hansen and M. Sekine. Decision diagram based techniques for the Haar wavelet transform. In *International Conference on Information, Communication & Signal Processing*, pages 59–63, 1997.

[84] M. A. Harrison. *Introduction to Switching and Automata Theory*. McGraw-Hill, 1965.

[85] K. D. Heidtmann. Arithmetic spectrum applied to fault detection for combinational networks. *IEEE Trans. on Comp.*, 40(3), 1991.

[86] K. W. Henderson. Comment on 'computation of the fast Walsh-Fourier transform'. *IEEE Trans. on Comp.*, 19:850–851, 1970.

[87] F. Hill and G. R. Peterson. *Computer Aided Logical Design with Emphasis on VLSI*. John Wiley & Sons, Inc., 1993.

[88] A. J. Hu. Formal hardware verification with BDDs: An introduction. In *Pacific Rim Conference on Communications, Computers and Signal Processing*, pages 677–682, 1997.

[89] S. L. Hurst. The application of Chow parameters and Rademacher-Walsh matrices in the synthesis of binary functions. *Comput. J.*, 16, no. 2, 1973.

[90] S. L. Hurst. *The Logical Processing of Digital Signals*. Crane-Russack, New York, 1978.

[91] S.L. Hurst. The Haar transform in digital network synthesis. In *Int'l Symp. on Multi-Valued Logic*, pages 10–18, 1981.

[92] S.L. Hurst, D.M.Miller, and J.C.Muzio. *Spectral Techniques in Digital Logic*. Academic Press Publishers, 1985.

[93] N. Ishiura, H. Sawada, and S. Yajima. Minimization of binary decision diagrams based on exchange of variables. In *Int'l Conf. on CAD*, pages 472–475, 1991.

[94] J. Jain, J. Bitner, D. Fussell, and J. Abraham. Probabilistic verification of Boolean functions. *Formal Methods in System Design: An International Journal*, 1(1):63–118, 1992.

[95] S.-W. Jeong, T.-S. Kim, and F. Somenzi. An efficient method for optimal BDD ordering computation. In *International Conference on VLSI and CAD*, 1993.

[96] B. Kapoor. Technology mapping and local optimization. In *European Design & Test Conf.*, pages 86–90, 1995.

[97] M. Karpovsky. *Finite Orthogonal Series in the Design of Digital Devices*. Wiley and JUP, 1976.

[98] U. Kebschull, E. Schubert, and W. Rosenstiel. Multilevel logic synthesis based on functional decision diagrams. In *European Conf. on Design Automation*, pages 43–47, 1992.

[99] C. Kern and M. R. Greenstreet. Formal verification in hardware design: A survey. *ACM Trans. on Design Automation of Electronic Systems*, 4:1–67, 1999.

[100] K. Keutzer. Dagon: Technology binding and local optimization by DAG matching. In *Design Automation Conf.*, pages 341–347, 1987.

[101] Z. Kohavi. *Switching and Finite Automata Theory*. McGraw-Hill Book Company, 1978.

[102] T. Kozlowski. *Application of Exclusive-OR Logic in Technology Independent Logic Optimisation*. Department of Electrical and Electronic Engineering, University of Bristol, Ph.D. Thesis, 1996.

[103] R. Krieger. PLATO: A tool for computation of exact signal probabilities. In *VLSI Design Conf.*, pages 65–68, 1993.

[104] S. K. Kumar and M. A. Breuer. Probabilistic aspects of Boolean switching functions via a new transform. *Journal of the ACM*, 28(3):502–520, 1981.

[105] Y.-T. Lai, M. Pedram, and S.B.K. Vrudhula. EVBDD-based algorithms for integer linear programming, spectral transformation, and function decomposition. *IEEE Trans. on CAD*, 13(8):959–975, 1994.

[106] Y.-T. Lai and S. Sastry. Edge-valued binary decision diagrams for multi-level hierarchical verification. In *Design Automation Conf.*, pages 608–613, 1992.

[107] Y.-T. Lai, S. Sastry, and M. Pedram. Boolean matching using binary decision diagrams with applications to logic synthesis and verification. In *Int'l Conf. on Comp. Design*, pages 452–458, 1992.

[108] L. Lavagno, P. McGeer, A. Saldanha, and A.L. Sangiovanni-Vincentelli. Timed shannon circuits: A power-efficient design style and synthesis tool. In *Design Automation Conf.*, pages 254–260, 1995.

[109] C.Y. Lee. Representation of switching circuits by binary decision diagrams. *Bell System Technical Jour.*, 38:985–999, 1959.

[110] A. M. Lloyd. A consideration of orthogonal matrices, other than the Rademacher-Walsh types, for the synthesis of digital networks. *J. Electronics*, 47:205–212, 1979.

[111] A. M. Lloyd. Design of multiplexor universal-logic-module networks using spectral techniques. *IEE Proceedings*, 127:31–36, 1980.

[112] F. Maihot and G. De Micheli. Algorithms for technology mapping based on binary decision diagrams and on Boolean operations. *IEEE Trans. on CAD*, pages 599–620, 1993.

[113] S. Malik, A.R. Wang, R.K. Brayton, and A.L. Sangiovanni-Vincentelli. Logic verification using binary decision diagrams in a logic synthesis environment. In *Int'l Conf. on CAD*, pages 6–9, 1988.

[114] E. J. McCluskey. *Introduction to the Theory of Switching Circuits.* McGraw-Hill, 1965.

[115] C. Meinel, F. Somenzi, and T. Theobald. Linear sifting of decision diagrams. In *Design Automation Conf.*, pages 202–207, 1997.

[116] C. Meinel and T. Theobald. Local encoding transformations for optimizing OBDD-representations of finite state machines. In *FMCAD*, volume 1166, pages 404–418, 1996.

[117] D. M. Miller. Graph algorithms for the manipulation of Boolean functions and their spectra. In *Congressus Numerantium*, pages 177–199, Winnipeg, Canada, 1987.

[118] D. M. Miller. A spectral method for Boolean function matching. In *European Design & Test Conf.*, page 602, 1996.

[119] D. M. Miller. An improved method for computing a generalized spectral coefficient. *IEEE Trans. on CAD*, 17:233–238, 1998.

[120] J. Mohnke and S. Malik. Permutation and phase independent Boolean comparison. In *European Conf. on Design Automation*, pages 86–92, 1993.

[121] J. Mohnke, P. Molitor, and S. Malik. Limits of using signatures for permutation independent Boolean comparison. In *ASP Design Automation Conf.*, pages 459–464, 1995.

[122] D. Möller, P. Molitor, and R. Drechsler. Symmetry based variable ordering for ROBDDs. *IFIP Workshop on Logic and Architecture Synthesis, Grenoble*, pages 47–53, 1994.

[123] D. E. Muller. Application of Boolean algebra to switching circuit design and error detection. *IRE Transactions*, 1:6–12, 1954.

[124] S. Muroga, Y. Kambayashi, H. C. Lai, and J. N. Culliney. The transduction method - design of logic networks based on permissible functions. *IEEE Trans. on Comp.*, 38, no. 10:1404–1424, 1989.

[125] J. Muzio and S. L. Hurst. The computation of complete and reduced sets of orthogonal spectral coefficients for logic design and pattern recognition purposes. *Computers and Electrical Engineering*, 5:231–249, 1978.

[126] S. Panda and F. Somenzi. Who are the variables in your neighborhood. In *Int'l Workshop on Logic Synth.*, pages 5b:5.11–5.20, 1995.

[127] S. Panda, F. Somenzi, and B.F. Plessier. Symmetry detection and dynamic variable ordering of decision diagrams. In *Int'l Conf. on CAD*, pages 628–631, 1994.

[128] K.P. Parker and E.J. McCluskey. Analysis of logic circuits with faults using input signal probabilities. *IEEE Trans. on Comp.*, 24:573–578, 1975.

[129] M. Pedram. Power minimization in IC design: Principles and applications. *ACM Trans. on Design Automation of Electronic Systems*, 1-1, 1996.

[130] P. D. Picton. Realisation of multithreshold threshold logic networks using the Rademacher-Walsh transform. *IEE Proceedings*, 128, pt. E, no. 3:107–113, 1981.

[131] S. Purwar and A. K. Susskind. Computation of Walsh spectrum from binary decision diagram and binary decision diagram from Walsh spectrum. *Computers and Electrical Engineering*, 15, no. 2:59–65, 1989.

[132] H. Rademacher. Einige Sätze über Reihen von allgemeinen orthogonal Funktionen. *Math. Ann.*, 87:112–138, 1922.

[133] I.S. Reed. A class of multiple-error-correcting codes and their decoding scheme. *IRE Trans. on Inf. Theory*, 3:6–12, 1954.

[134] D.E. Ross, K.M. Butler, R. Kapur, and M.R. Mercer. Fast functional evaluation of candidate OBDD variable ordering. In *European Conf. on Design Automation*, pages 4–9, 1991.

[135] R. Rudell. Dynamic variable ordering for ordered binary decision diagrams. In *Int'l Conf. on CAD*, pages 42–47, 1993.

[136] R. Rudell and A. L. Sangiovanni-Vincintelli. Espresso-MV: Algorithms for multiple-valued logic minimization. In *IEEE Custom Integrated Circuit Conference*, pages 230–234, Portland, Oregon, 1985.

[137] T. Sasao. *Logic Synthesis and Optimization*. Kluwer Academic Publisher, 1993.

[138] T. Sasao. Representations of logic functions by using EXOR operators. In T. Sasao and M. Fujita, editors, *Representation of Discrete Functions*, pages 29–54. Kluwer Academic Publishers, 1996.

[139] T. Sasao. *Switching Theory for Logic Synthesis*. Kluwer Academic Publishers, 1999.

[140] T. Sasao and J. Butler. A design method for look-up table type FPGA by pseudo-Kronecker expansion. In *Int'l Symp. on Multi-Valued Logic*, pages 97–106, 1994.

[141] H. Savoj, M. J. Silva, R. K. Brayton, and A. Sangiovani-Vincentelli. Boolean matching in logic synthesis. In *European Design & Test Conf.*, pages 168–174, 1992.

[142] U. Schlichtmann and F. Brglez. Efficient Boolean matching in technology mapping with very large cell libraries. In *Custom Integrated Circuits Conference*, pages 3.6.1–3.6.6, 1993.

[143] U. Schlichtmann, F. Brglez, and P. Schneider. Efficient Boolean matching based on unique variable ordering. In *Int'l Workshop on Logic Synth.*, pages 3b1–3b13, 1993.

[144] C. Scholl and B. Becker. On the generation of multiplexer circuits for pass transistor logic. In *Design, Automation and Test in Europe*, pages 372–378, 2000.

[145] E. Sentovich, K. Singh, L. Lavagno, Ch. Moon, R. Murgai, A. Saldanha, H. Savoj, P. Stephan, R. Brayton, and A. Sangiovanni-Vincentelli. SIS: A system for sequential circuit synthesis. Technical report, University of Berkeley, 1992.

[146] J. L. Shanks. Computation of the fast Walsh-Fourier transform. *IEEE Trans. on Comp.*, 18:457–459, 1969.

[147] C. E. Shannon. Symbolic analysis of relay and switching circuits. *Transactions of the AIEE*, 57:713–723, 1938.

[148] F. Somenzi. *CUDD: CU Decision Diagram Package Release 2.1.2*. University of Colorado at Boulder, 1997.

[149] R. S. Stankovic. Some remarks about spectral transform interpretation of MTBDDs and EVBDDs. In *ASP Design Automation Conf.*, pages 385–390, 1995.

[150] R. S. Stankovic, T. Sasao, and C. Moraga. Spectral transforms decision diagrams. In T. Sasao and M. Fujita, editors, *Representation of Discrete Functions*, pages 55–92. Kluwer Academic Publishers, 1996.

[151] P. Tafertshofer and M. Pedram. Factored edge-valued binary decision diagrams. *Formal Methods in System Design: An International Journal*, 10(2):243–270, 1997.

[152] M. A. Thornton. Modified Haar transform calculation using digital circuit output probabilities. In *International Conference on Information, Communication & Signal Processing*, pages 52–58, 1997.

[153] M. A. Thornton and R. Drechsler. Spectral decision diagrams using graph transformations. In *Design, Automation and Test in Europe*, 2001.

[154] M. A. Thornton, R. Drechsler, and W. Günther. A method for approximate equivalence checking. In *Int'l Symp. on Multi-Valued Logic*, pages 447,452, 2000.

[155] M. A. Thornton and V. S. S. Nair. Combinational logic synthesis using spectral techniques. In *European Design Automation Conf.*, pages 358–363, 1993.

[156] M. A. Thornton and V. S. S. Nair. Behavioral to structural translation in ESOP form using the verilog HDL. In *International Verilog HDL Conference*, pages 58–62, 1994.

[157] M. A. Thornton and V. S. S. Nair. Efficient calculation of spectral coefficients and their application. *IEEE Trans. on CAD*, 14(11):1328–1341, 1995.

[158] M. A. Thornton and V. S. S. Nair. Fast Reed-Muller spectrum computation using output probabilities. In *Workshop on Applications of the Reed-Muller Expansion in Circuit Design*, pages 281–287, Makuhari, Chiba, Japan, 1995.

[159] M. A. Thornton and V. S. S. Nair. Parity function detection and realization using a small set of spectral coefficients. In *Int'l Workshop on Logic Synth.*, Tahoe City, California, 1995.

[160] M. A. Thornton and V. S. S. Nair. BDD based spectral approach for Reed-Muller circuit realisation. *IEE Proceedings*, 193(2):145–150, 1996.

[161] M. A. Thornton and V. S. S. Nair. Behavioral synthesis of combinational logic using spectral based heuristics. *ACM Trans. on Design Automation of Electronic Systems*, 4-2:219–230, 1999.

[162] M. A. Thornton, J. Williams, R. Drechsler, and N. Drechsler. Variable reordering for shared binary decision diagrams using output probabilities. In *Design, Automation and Test in Europe*, pages 758–759, 1999.

[163] C. Y. Tsui, M. Pedram, and A. Despain. Efficient estimation of dynamic power dissipation under a real delay model. In *Int'l Conf. on CAD*, pages 224–228, 1993.

[164] D. Varma and E. A. Trachtenberg. Design automation tools for efficient implementation of logicfunctions by decomposition. *IEEE Trans. on CAD*, 8:901–916, 1989.

[165] S.B.K. Vrudhula, M. Pedram, and Y.-T. Lai. Edge valued binary decision diagrams. In T. Sasao and M. Fujita, editors, *Representation of Discrete Functions*, pages 109–132. Kluwer Academic Publisher, 1996.

[166] J. S. Wallis. *Hadamard Matrices*. (Lecture Notes No. 292), Springer-Verlag, 1972.

[167] J. L. Walsh. A closed set of normal orthogonal functions. *American Journal of Mathematics*, 55:5–24, 1923.

[168] X. Wang. *Technology Mapping for FPGA's using Boolean Matching and Spectral Techniques*. M.S. Thesis, University of Victoria, 1996.

[169] A. T. White. *Graphs, Groups and Surfaces*. North-Holland Publishing Company, 1973.

[170] J. C.-Y. Yang and G. De Micheli. Spectral techniques for technology mapping. Technical report, Technical Report CSL-TR-91-498, Stanford University, 1991.

[171] K. Yano, Y. Sasaki, K. Rikino, and K. Seki. Top-down pass-transistor logic design. *IEEE Jour. of Solid-State Circ.*, 31(6):792–803, June 1996.

[172] K. Zhu and D. F. Wong. Fast Boolean matching for FPGAs. In *European Conf. on Design Automation*, pages 352–357, 1993.

Index